TSA Practice Papers

Volume One

UniAdmissions

Copyright © 2018 *UniAdmissions*. All rights reserved.

ISBN 978-1-912557-43-1

No part of this publication may be reproduced or transmitted in any form or by any means, electronic or mechanical, including photocopying, recording, or by any information retrieval system without prior written permission of the publisher. This publication may not be used in conjunction with or to support any commercial undertaking without the prior written permission of the publisher.

Published by *RAR Medical Services Limited*
www.uniadmissions.co.uk
info@uniadmissions.co.uk
Tel: 0208 068 0438

This book is neither created nor endorsed by TSA. The authors and publisher are not affiliated with TSA. The information offered in this book is purely advisory and any advice given should be taken within this context. As such, the publishers and authors accept no liability whatsoever for the outcome of any applicant's TSA performance, the outcome of any university applications or for any other loss. Although every precaution has been taken in the preparation of this book, the publisher and author assume no responsibility for errors or omissions of any kind. Neither is any liability assumed for damages resulting from the use of information contained herein. This does not affect your statutory rights.

TSA Practice Papers

3 Full Papers & Solutions

Volume One

Rohan Agarwal
Jonathan Madigan

UniAdmissions

About the Authors

Jon studied Economics and Management at St Hugh's College, Oxford, between 2013 and 2016. He sat the Thinkings Skills Assessment and **scored full marks** in section 1 of the paper, placing him in the **top 0.1% of candidates** who sat the assessment that year.

Jon has worked with UniAdmissions since 2014, working primarily as a personal tutor for a number of applicants across the entire application process. He has also provided Thinking Skills Assessment preparation courses within schools, and remains very familiar with all aspects of the paper.

Although Jon has moved into finance since graduation, he remains involved with tutoring as much as time allows. He regularly visits Oxford in his free time, and hopes to return for further study at some point in the future.

Rohan is the **Director of Operations** at *UniAdmissions* and is responsible for its technical and commercial arms. He graduated from Gonville and Caius College, Cambridge and is a fully qualified doctor. Over the last five years, he has tutored hundreds of successful Oxbridge and Medical applicants. He has also authored ten books on admissions tests and interviews.

Rohan has taught physiology to undergraduates and interviewed medical school applicants for Cambridge. He has published research on bone physiology and writes education articles for the Independent and Huffington Post. In his spare time, Rohan enjoys playing the piano and table tennis.

Introduction 6
General Advice 8
Revision Timetable 13
Getting the most out of Mock Papers 14
Before using this Book 15
Section 1: An Overview 17
Section 2: An Overview 18
Section 2: Revision Guide 19
Scoring Tables 22

Mock Papers 23
Mock Paper A 23
Mock Paper B 34
Mock Paper C 47

Answer Key 66

Mock Paper Answers 67
Mock Paper A: Section 1 67
Mock Paper A: Section 2 72
Mock Paper B: Section 1 76
Mock Paper B: Section 2 82
Mock Paper C: Section 1 86
Mock Paper C: Section 2 94

Final Advice 98

Introduction

The Basics

The Thinking Skills Assessment is an aptitude test taken by students who are applying to certain courses at Cambridge and Oxford. Cambridge applicants sit the TSA Cambridge and Oxford applicants sit the TSA Oxford.

SECTION	SKILLS TESTED	QUESTIONS	TIMING
ONE	Problem-solving skills, including numerical and spatial reasoning. Critical thinking skills, including understanding argument and reasoning using everyday language.	50 MCQs	90 minutes
TWO	Ability to organise ideas in a clear and concise manner, and communicate them effectively in writing. Questions are usually but not necessarily medical.	One Essay from Four	30 minutes

NB: **TSA Oxford** consists of sections 1 + 2; **TSA Cambridge** and **TSA UCL** consist of only section 1.

Who has to sit the TSA?

Exam	Course
TSA Oxford	Students applying for the following subjects at Oxford **MUST** take the TSA: Economics and Management; Experimental Psychology; History and Economics; Human Sciences; Philosophy and Linguistics; Philosophy, Politics and Economics (PPE); Psychology and Linguistics; Psychology and Philosophy
TSA Cambridge	Students applying for Land Economy **MAY** have to take the TSA depending on the college they apply to.
TSA UCL	Students applying for European Social and Political Students **MUST** take the TSA.

NB: Applicants for Oxford **Chemistry or History & Economics only have to complete section 1 of the TSA.**

Preparing for the TSA

Before going any further, it's important that you understand the optimal way to prepare for the TSA. Rather than jumping straight into doing mock papers, it's essential that you start by understanding the components and the theory behind the TSA by using a TSA textbook. Once you've finished the non-timed practice questions, you can progress to past TSA papers. These are freely available online at www.uniadmissions.co.uk/tsa-past-papers and serve as excellent practice. You're strongly advised to use these in combination with the *TSA Past Worked Solutions* Book so that you can improve your weaknesses. Finally, once you've exhausted past papers, move onto the mock papers in this book.

Learn TSA Theory & Techniques
- Ultimate TSA Guide

→ Practice Questions
- TSA Past Paper Worked Solutions

→ TSA Mock Papers
- TSA Practice Papers

Already seen them all?

So, you've run out of past papers? Well hopefully that is where this book comes in. It contains six unique mock papers; each compiled by TSA Expert tutors at *UniAdmissions* and available nowhere else.

Having successfully gained a place at Oxford and scoring in the top 10% of the TSA, our tutors are intimately familiar with the TSA and its associated admission procedures. So, the novel questions presented to you here are of the correct style and difficulty to continue your revision and stretch you to meet the demands of the TSA.

General Advice

Start Early

It is much easier to prepare if you practice little and often. Start your preparation well in advance; ideally 10 weeks but at the latest within a month. This way you will have plenty of time to complete as many papers as you wish to feel comfortable and won't have to panic and cram just before the test, which is a much less effective and more stressful way to learn. In general, an early start will give you the opportunity to identify the complex issues and work at your own pace.

Prioritise

The MCQ section can be very time-pressured, and if you fail to answer the questions within the time limit you will be doing yourself a major disservice as every mark counts for this section. You need to be aware of how much time you're spending on each passage and allocate your time wisely. For example, since there are 50 questions in Section 1, and you are given 90 minutes in total, you will ideally take about 100 seconds per question (including reading time) so that you will not run out of time and panic towards the end.

Positive Marking

There are no penalties for incorrect answers; you will gain one for each right answer and will not get one for each wrong or unanswered one. This provides you with the luxury that you can always guess should you absolutely be not able to figure out the right answer for a question or run behind time. Since each question in Section 1 provides you with 5 possible answers, you have a 20% chance of guessing correctly. Therefore, if you aren't sure (and are running short of time), then make an educated guess and move on. Before 'guessing' you should try to eliminate a couple of answers to increase your chances of getting the question correct. For example, if a question has 5 options and you manage to eliminate 2 options- your chances of getting the question increase from 20% to 33%!

Avoid losing easy marks on other questions because of poor exam technique. Similarly, if you have failed to finish the exam, take the last ten seconds to guess the remaining questions to at least give yourself a chance of getting them right.

Practice

This is the best way of familiarising yourself with the style of questions and the timing for this section. Although the test does not demand any technical knowledge, you are unlikely to be familiar with the style of questions in all sections when you first encounter them. Therefore, you want to be comfortable at using this before you sit the test.

Practising questions will put you at ease and make you more comfortable with the exam. The more comfortable you are, the less you will panic on the test day and the more likely you are to score highly. Initially, work through the questions at your own pace, and spend time carefully reading the questions and looking at any additional data. When it becomes closer to the test, **make sure you practice the questions under exam conditions**.

Past Papers

Official past papers and answers are freely available online at **www.uniadmissions.co.uk/tsa-past-papers**. Practice makes perfect, and the more you practice the questions, especially for Section 1, the better you will get. Do not worry if you make plenty of mistakes at the start, the best way to learn is to understand why you have made certain mistakes and to not commit them again in the future!

You will undoubtedly get stuck when doing some past paper questions – they are designed to be tricky and the answer schemes don't offer any explanations. Thus, **you're highly advised to acquire a copy of *TSA Past Paper Worked Solutions*** – a free ebook is available online (see the back of this book for more details).

Repeat Questions

When checking through answers, pay particular attention to questions you have got wrong. If there is a worked answer, look through that carefully until you feel confident that you understand the reasoning, and then repeat the question without help to check that you can do it. If only the answer is given, have another look at the question and try to work out why that answer is correct. This is the best way to learn from your mistakes, and means you are less likely to make similar mistakes when it comes to the test. The same applies for questions which you were unsure of and made an educated guess which was correct, even if you got it right. When working through this book, **make sure you highlight any questions you are unsure of**, this means you know to spend more time looking over them once marked.

No Calculators and Dictionaries

The TSA requires a strong command of the English language, especially for Section 2 where you are asked to write an essay in 30 minutes. You are not allowed to use spell check or a dictionary, hence you should ensure that you written English is up to standard and you should ideally make close to no grammatical or spelling errors for your essay.

Section 1 contains several numerical reasoning questions, and you are not allowed to use a calculator, so make sure you are careful with your calculations.

Keywords

If you're stuck on a question, sometimes you can simply quickly scan the passage for any keywords that match the questions.

A word on Timing...

"If you had all day to do your exam, you would get 100%. But you don't."

Whilst this isn't completely true, it illustrates a very important point. Once you've practiced and know how to answer the questions, the clock is your biggest enemy. This seemingly obvious statement has one very important consequence. **The way to improve your score is to improve your speed.** There is no magic bullet. But there are a great number of techniques that, with practice, will give you significant time gains, allowing you to answer more questions and score more marks.

Timing is tight throughout – **mastering timing is the first key to success**. Some candidates choose to work as quickly as possible to save up time at the end to check back, but this is generally not the best way to do it. Often questions can have a lot of information in them – each time you start answering a question it takes time to get familiar with the instructions and information. By splitting the question into two sessions (the first run-through and the return-to-check) you double the amount of time you spend on familiarising yourself with the data, as you have to do it twice instead of only once. This costs valuable time. In addition, candidates who do check back may spend 2–3 minutes doing so and yet not make any actual changes. Whilst this can be reassuring, it is a false reassurance as it is unlikely to have a significant effect on your actual score. Therefore, it is usually best to pace yourself very steadily, aiming to spend the same amount of time on each question and finish the final question in a section just as time runs out. This reduces the time spent on re-familiarising with questions and maximises the time spent on the first attempt, gaining more marks.

It is essential that you don't get stuck with the hardest questions – no doubt there will be some. In the time spent answering only one of these you may miss out on answering three easier questions. If a question is taking too long, choose a sensible answer and move on. Never see this as giving up or in any way failing, rather it is the smart way to approach a test with a tight time limit. With practice and discipline, you can get very good at this and learn to maximise your efficiency. It is not about being a hero and aiming for full marks – this is almost impossible and very much unnecessary. It is about maximising your efficiency and gaining the maximum possible number of marks within the time you have.

Use the Options:

Some passages may try to trick you by providing a lot of unnecessary information. When presented with long passages that are seemingly hard to understand, it's essential you look at the answer options so you can focus your mind. This can allow you to reach the correct answer a lot more quickly. Consider the example below:

'Mountain climbing is viewed by some as an extreme sport, while for others it is simply an exhilarating pastime that offers the ultimate challenge of strength, endurance, and sacrifice. It can be highly dangerous, even fatal, especially when the climber is out of his or her depth, or simply gets overwhelmed by weather, terrain, ice, or other dangers of the mountain. Inexperience, poor planning, and inadequate equipment can all contribute to injury or death, so knowing what to do right matters.

Despite all the negatives, when done right, mountain climbing is an exciting, exhilarating, and rewarding experience. This article is an overview beginner's guide and outlines the initial basics to learn. Each step is deserving of an article in its own right, and entire tomes have been written on climbing mountains, so you're advised to spend a good deal of your beginner's learning immersed in reading widely. This basic overview will give you an idea of what is involved in a climb.'

Looking at the options first makes it obvious that certain information is redundant and allows you to quickly zoom in on certain keywords you should pick up on in order to answer the questions.

In other cases, **you may actually be able to solve the question without having to read the passage over and over again**. For example:

Which statement best summarises this paragraph?
A. Mountain climbing is an extreme sport fraught with dangers.
B. Without extensive experience embarking on a mountain climb is fatal.
C. A comprehensive literature search is the key to enjoying mountain climbing.
D. Mountain climbing is difficult and is a skill that matures with age if pursued.
E. The terrain is the biggest unknown when climbing a mountain and therefore presents the biggest danger.

If you read the passage first before looking at the question, you might have forgotten what the passage mentioned, and you will have to spend extra time going back to the passage to re-read it again.

You can **save a lot of time by looking at the questions first before reading the passage**. After looking at the question, you will know at the back of your head to look out for and this will save a considerable amount of time.

Manage your Time:

It is highly likely that you will be juggling your revision alongside your normal school studies. Whilst it is tempting to put your A-levels on the back burner falling behind in your school subjects is not a good idea, don't forget that to meet the conditions of your offer should you get one you will need at least one A*. So, time management is key!

Make sure you set aside a dedicated 90 minutes (and much more closer to the exam) to commit to your revision each day. The key here is not to sacrifice too many of your extracurricular activities, everybody needs some down time, but instead to be efficient. Take a look at our list of top tips for increasing revision efficiency below:

1. Create a comfortable work station
2. Declutter and stay tidy
3. Treat yourself to some nice stationery
4. See if music works for you → if not, find somewhere peaceful and quiet to work
5. Turn off your mobile or at least put it into silent mode
6. Silence social media alerts
7. Keep the TV off and out of sight
8. Stay organised with to do lists and revision timetables – more importantly, stick to them!
9. Keep to your set study times and don't bite off more than you can chew
10. Study while you're commuting
11. Adopt a positive mental attitude
12. Get into a routine
13. Consider forming a study group to focus on the harder exam concepts
14. Plan rest and reward days into your timetable – these are excellent incentive for you to stay on track with your study plans!

Keep Fit & Eat Well:

'A car won't work if you fill it with the wrong fuel' - your body is exactly the same. You cannot hope to perform unless you remain fit and well. The best way to do this is not underestimate the importance of healthy eating. Beige, starchy foods will make you sluggish; instead start the day with a hearty breakfast like porridge. Aim for the recommended 'five a day' intake of fruit/veg and stock up on the oily fish or blueberries – the so called "super foods".

When hitting the books, it's essential to keep your brain hydrated. If you get dehydrated you'll find yourself lethargic and possibly developing a headache, neither of which will do any favours for your revision. Invest in a good water bottle that you know the total volume of and keep sipping throughout the day. Don't forget that the amount of water you should be aiming to drink varies depending on your mass, so calculate your own personal recommended intake as follows: 30 ml per kg per day.

It is well known that exercise boosts your wellbeing and instils a sense of discipline. All of which will reflect well in your revision. It's well worth devoting half an hour a day to some exercise, get your heart rate up, break a sweat, and get those endorphins flowing.

Sleep

It's no secret that when revising you need to keep well rested. Don't be tempted to stay up late revising as sleep actually plays an important part in consolidating long term memory. Instead aim for a minimum of 7 hours good sleep each night, in a dark room without any glow from electronic appliances. Install flux (https://justgetflux.com) on your laptop to prevent your computer from disrupting your circadian rhythm. Aim to go to bed the same time each night and no hitting snooze on the alarm clock in the morning!

Revision Timetable

Still struggling to get organised? Then try filling in the example revision timetable below, remember to factor in enough time for short breaks, and stick to it! Remember to schedule in several breaks throughout the day and actually use them to do something you enjoy e.g. TV, reading, YouTube etc.

	10AM	12PM	2PM	4PM	8PM
TUESDAY					
WEDNESDAY					
THURSDAY					
FRIDAY					
SATURDAY					
SUNDAY					
EXAMPLE DAY	School			Critical Thinking	Essay

you have a much more accurate idea of the time you're spending on a question. In general, if you've spent seconds on a section 1 question – move on regardless of how close you think you are to solving it.

Getting the most out of Mock Papers

Mock exams can prove invaluable if tackled correctly. Not only do they encourage you to start revision earlier, they also allow you to **practice and perfect your revision technique**. They are often the best way of improving your knowledge base or reinforcing what you have learnt. Probably the best reason for attempting mock papers is to familiarise yourself with the exam conditions of the TSA as they are particularly tough.

Start Revision Earlier
Thirty five percent of students agree that they procrastinate to a degree that is detrimental to their exam performance. This is partly explained by the fact that they often seem a long way in the future. In the scientific literature this is well recognised, Dr. Piers Steel, an expert on the field of motivation states that *'the further away an event is, the less impact it has on your decisions'*.

Mock exams are therefore a way of giving you a target to work towards and motivate you in the run up to the real thing – every time you do one treat it as the real deal! If you do well then it's a reassuring sign; if you do poorly then it will motivate you to work harder (and earlier!).

Practice and perfect revision techniques
In case you haven't realised already, revision is a skill all to itself, and can take some time to learn. For example, the most common revision techniques including **highlighting and/or re-reading are quite ineffective** ways of committing things to memory. Unless you are thinking critically about something you are much less likely to remember it or indeed understand it.

Mock exams, therefore allow you to test your revision strategies as you go along. Try spacing out your revision sessions so you have time to forget what you have learnt in-between. This may sound counterintuitive but the second time you remember it for longer. Try teaching another student what you have learnt; this forces you to structure the information in a logical way that may aid memory. Always try to question what you have learnt and appraise its validity. Not only does this aid memory but it is also a useful skill for the TSA, Oxbridge interviews, and beyond.

Improve your knowledge
The act of applying what you have learnt reinforces that piece of knowledge. An essay question in Section 2 may ask you about a fairly simple topic, but if you have a deep understanding of it you are able to write a critical essay that stands out from the crowd. Essay questions in particular provide a lot of room for students who have done their research to stand out, hence you should always aim to improve your knowledge and apply it from time to time. As you go through the mocks or past papers take note of your performance and see if you consistently under-perform in specific areas, thus highlighting areas for future study.

Get familiar with exam conditions
Pressure can cause all sorts of trouble for even the most brilliant students. The TSA is a particularly time pressured exam with high stakes – your future (without exaggerating) does depend on your result to a great extent. The real key to the TSA is overcoming this pressure and remaining calm to allow you to think efficiently.

Mock exams are therefore an excellent opportunity to devise and perfect your own exam techniques to beat the pressure and meet the demands of the exam. **Don't treat mock exams like practice questions – it's imperative you do them under time conditions.**

Before using this Book

Do the ground work
- Understand the format of the TSA – have a look at the TSA website and familiarise yourself with it: www.admissionstesting.org/thinking-skills-assessment
- Read widely in order to prepare yourself for the essay.
- Improve your written English if you are not confident in this aspect by practicing writing and reading frequently.
- Try to broaden your reading by learning about different topics that you are unfamiliar with as the essay topics can vary greatly.
- Learn how to understand a writer's viewpoint by reading news articles and having a go at summarising what the writer is arguing about.
- Be consistent – slot in regular TSA practice sessions when you have pockets of free time.
- Engage in discussion sessions with your friends and teachers – this might give you more ideas about certain essay topics.

Ease in gently
With the ground work laid, there's still no point in adopting exam conditions straight away. Instead invest in a beginner's guide to the TSA, which will not only describe in detail the background and theory of the exam, but take you through section by section what is expected. *The Ultimate TSA Guide* is the most popular TSA textbook – you can get a free copy by flicking to the back of this book.

When you are ready to move on to past papers, take your time and puzzle your way through all the questions. Really try to understand solutions. A past paper question won't be repeated in your real exam, so don't rote learn methods or facts. Instead, focus on applying prior knowledge to formulate your own approach.

If you're really struggling and have to take a sneak peek at the answers, then practice thinking of alternative solutions, or arguments for essays. It is unlikely that your answer will be more elegant or succinct than the model answer, but it is still a good task for encouraging creativity with your thinking. Get used to thinking outside the box!

Accelerate and Intensify

Start adopting exam conditions after you've done two past papers. Don't forget that **it's the time pressure that makes the TSA hard** – if you had as long as you wanted to sit the exam you would probably get 100%. If you're struggling to find comprehensive answers to past papers then *TSA Past Papers Worked Solutions* contains detailed explained answers to every TSA past paper question and essay (flick to the back to get a free copy).

Doing all the past papers is a good target for your revision. Choose a paper and proceed with strict exam conditions. Take a short break and then mark your answers before reviewing your progress. For revision purposes, as you go along, keep track of those questions that you guess – these are equally as important to review as those you get wrong.

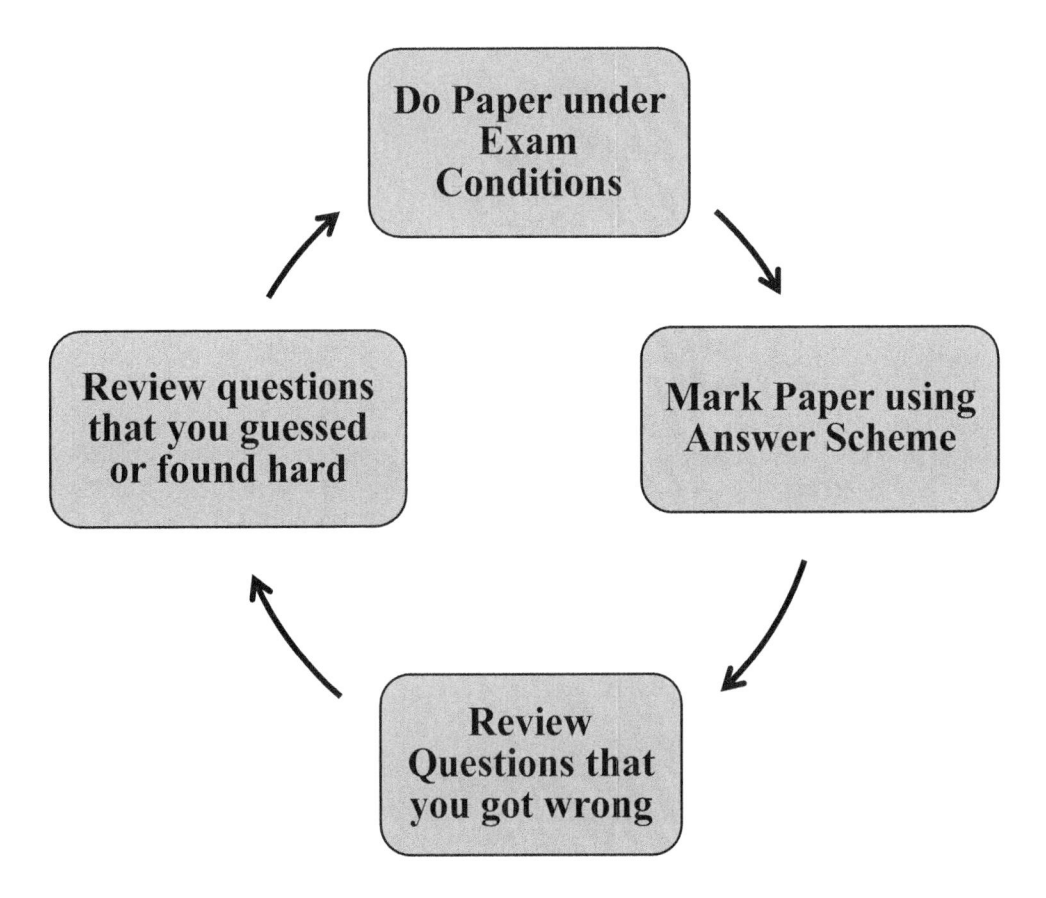

```
        Do Paper under
         Exam Conditions
              ↓
Review questions        Mark Paper using
that you guessed   ←    Answer Scheme
or found hard           
              ↑              ↓
          Review Questions that
            you got wrong
```

Once you've exhausted all the past papers, move on to tackling the unique mock papers in this book. In general, you should aim to complete one to two mock papers every night in the ten days preceding your exam.

Section 1: An Overview

What will you be tested on?	No. of Questions	Duration
Problem-solving skills, numerical and spatial reasoning, critical thinking skills, understanding arguments and reasoning	50 MCQs	90 Minutes

This is the first section of the TSA, comprising a total of 50 MCQ questions. You have 90 minutes in total to complete the MCQ questions, including reading time. In order to keep within the time limit, you realistically have about 108 seconds per question as you have to factor in the reading time as well.

Not all the questions are of equal difficulty and so as you work through the past material it is certainly worth learning to recognise quickly which questions you should spend less time on in order to give yourself more time for the trickier questions.

Deducing arguments
Several MCQ questions will be aimed at testing your understand of the writer's argument. It is common to see questions asking you 'what is the writer's view?' or 'what is the writer trying to argue?'. This is arguably an important skill you will have to develop, and the TSA is designed to test this ability. You have limited time to read the passage and understand the writer's argument, and the only way to improve your reading comprehension skill is to read several well-written news articles on a daily basis and think about them in a critical manner.

Assumptions
It is important to be able to identify the assumptions that a writer makes in the passage, as several questions might question your understand of what is assumed in the passage. For example, if a writer mentions that 'if all else remains the same, we can expect our economic growth to improve next year', you can identify an assumption being made here – the writer is clearly assuming that all external factors remain the same.

Fact vs. Opinion
It is important to **be able to decipher whether the writer is stating a fact or an opinion** – the distinction is usually rather subtle and you will have to decide whether the writer is giving his or her own personal opinion, or presenting something as a fact. Section 1 may contain questions that will test your ability to identify what is presented as a fact and what is presented as an opinion.

Fact	Opinion
'There are 7 billion people in this world…'	'I believe there are more than 7 billion people in this world…'
'She is an Australian…'	'She sounded like an Australian…'
'Trump is the current President…'	'Trump is a horrible President…'
'Vegetables contain a lot of fibre…'	'Vegetables are good for you…'

Numerical and spatial reasoning
There are several questions that will test how well you can cope with numbers, and you should ideally be comfortable with simple mental calculations and being able to think logically.

Section 2: An Overview

What will you be tested on?	No. of Questions	Duration
Your ability to write an essay under timed conditions, your writing technique and your argumentative abilities	1 out of 4	30 Minutes

Section 2 is usually what students are more comfortable with – after all, many GCSE and A Level subjects require you to write essays within timed conditions. This section does not require you to have any particular specialist knowledge – the questions can be very broad and cover a wide range of topics.

Here are some of the topics that might appear in Section 2:

- Science
- Politics
- Religion
- Technology
- Ethics
- Morality
- Philosophy
- Education
- History
- Geopolitics

As you can see, this list is very broad and definitely non-exhaustive, and you do not get many choices to choose from (you have to write one essay out of three choices). Many students make the mistake of focusing too narrowly on one or two topics that they are comfortable with – this is a dangerous gamble and if you end up with four questions you are unfamiliar with, this is likely to negatively impact your score. **You should ideally focus on three topics to prepare from the list above**, and you can pick and choose which topics from the list above are the ones you would be more interested in. Here are some suggestions:

Science
An essay that is related to science might relate to recent technological advancements and their implications, such as the rise of Bitcoin and the use of blockchain technology and artificial intelligence. This is interrelated to ethical and moral issues, hence you cannot merely just regurgitate what you know about artificial intelligence or blockchain technology. **The examiners do not expect you to be an expert in an area of science** – what they want to see is how you identify certain moral or ethical issues that might arise due to scientific advancements, and how do we resolve such conundrums as human beings.

Politics
Politics is undeniably always a hot topic and consequently a popular choice amongst students. The danger with writing a politics question is that some **students get carried away and make their essay too one-sided or emotive** – for example a student may chance upon an essay question related to Brexit and go on a long rant about why the referendum was a bad idea. You should always remember to answer the question and make sure your essay addresses the exact question asked – do not get carried away and end up writing something irrelevant just because you have strong feelings about a certain topic.

Religion
Religion is always a controversial issue and essays on religion provide good students with an excellent opportunity to stand out and display their maturity in thought. Questions can range from asking about your opinion with regards to banning the wearing of a headdress to whether children should be exposed to religious practices at a young age. Questions related to religion will **require you to be sensitive and measured** in your answers - it is easy to trip up on such questions if you're not careful.

Education
Education is perhaps always a relatable topic to students, and students can draw from their own experience with the education system in order to form their opinion and write good essays on such topics. Questions can range from whether university places should be reduced, to whether we should be focusing on learning the sciences as opposed to the arts.

Section 2: Revision Guide

Science

Resource	What to read/do
1. Newspaper Articles	• The Guardian, The Times, The Economist, The Financial Times, The Telegraph, The New York Times, The Independent
2. A Levels/IB	• Look at the content of your science A Levels/IB if you are doing science subjects and critically analyse what are the potential moral/ethical implications • Use your A Levels/IB resources in order to seek out further readings – e.g. links to a scientific journal or blog commentary • Remember that for your LNAT essay you should not focus on the technical issues too much – think more about the ethical and moral issues
3. Online videos	• There are plenty of free resources online that provide interesting commentary on science and the moral and ethical conundrums that scientists face on a daily basis • E.g. Documentaries and specialist science channels on YouTube • National Geographic, Animal Planet etc. might also be good if you have access to them
4. Debates	• Having a discussion with your friends about topics related to science might also help you formulate some ideas • Attending debate sessions where the topic is related to science might also provide you with excellent arguments and counter-arguments • Some universities might also host information sessions for sixth form students – some might be relevant to ethical and moral issues in science
5. Museums	• Certain museums such as the Natural Science Museum might provide some interesting information that you might not have known about
6. Non-fiction books	• There are plenty of non-fiction books (non-technical ones) that might discuss moral and ethical issues about science in an easily digestible way

Politics

Resource	What to read/do
1. Newspaper Articles	• The Guardian, The Times, The Economist, The Financial Times, The Telegraph, The New York Times, The Independent
2. Television	• Parliamentary sessions • Prime Minister Questions • Political news
3. Online videos	• Documentaries • YouTube Channels
4. Lectures	• University introductory lectures • Sixth form information sessions
5. Debates	• Debates held in school • Joining a politics club
6. Podcasts	• Political podcasts • Listen to both sides to get a more rounded view (e.g. listening to both left and right wing podcasts)

Religion

Syllabus Point	What to read/do
2. Non-fiction books	• Read up about books that explain the origins and beliefs of different type religion • E.g. Books that talk about the origins of Christianity, Islam or Buddh theology books etc.
3. Talking to religious leaders	religions more and being able to write an essay on religion with more mat and nuance • Talking to people from different religious backgrounds may also be a good of forming a more well-rounded opinion
5. Lectures	• Information sessions • Relevant introductory lectures

Education

	What to read/do
1. Newspaper Articles	• The Guardian, The Times, The Economist, The Financial Times, Telegraph, The New York Times, The Independent
2. A Levels/IB	feel like what you are studying is useful and relevant? E.g. Studying versus science • Compare the education you are receiving with your friends in differ schools or different subjects
4. University applications	• Have a read of how different universities promote themselves – do they cl to provide students with academic enlightenment, or better job prospects a good social life? • Why do different universities focus on different things?
6. Talk to your teachers	• Your teachers have been in the education industry for years and ma decades – talk to them and ask them for their opinion • Talk to different teachers and compare their opinions regarding how

rather than doing extra reading as the former will have a greater impact on your mark.

How to use this Book

If you have done everything this book has described so far then you should be well equipped to meet the demands of the TSA, and therefore **the mock papers in the rest of this book should ONLY be completed under exam conditions**.

This means:
- Absolute silence – no TV or music
- Absolute focus – no distractions such as eating your dinner
- Strict time constraints – no pausing half way through
- No checking the answers as you go
- Give yourself a maximum of three minutes between sections – keep the pressure up
- Complete the entire paper before marking
- Mark harshly

In practice this means setting aside 90 minutes for Section 1 and 30 minutes for Section 2 in an evening to find a quiet spot without interruptions and tackle the paper. Completing one mock paper every evening in the week running up to the exam would be an ideal target.

- Tackle the paper as you would in the exam.
- Return to mark your answers, but mark harshly if there's any ambiguity.
- Highlight any areas of concern.
- If warranted read up on the areas you felt you underperformed to reinforce your knowledge.
- If you inadvertently learnt anything new by muddling through a question, go and tell somebody about it to reinforce what you've discovered.

Finally relax… the TSA is an exhausting exam, concentrating so hard continually for 1.5 hours will take its toll. So, being able to relax and switch off is essential to keep yourself sharp for exam day! Make sure you reward yourself after you finish marking your exam.

Scoring Tables

Use these to keep a record of your scores from past papers – you can then easily see which paper you should attempt next (always the one with the lowest score).

	2nd Attempt	3rd Attempt
2008		
2010		
2012		
2015		

Top Tip! When repeating a mock paper, its best to attempt a different essay title to give yourself max experience with the various styles of TSA essays.

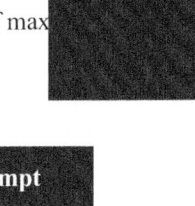

	SECTION 1	1st Attempt	2nd Attempt	3rd Attempt
Volume Two				
	Mock Paper D			

You will not be able to give yourself a score for Section 2– the best way to gauge your performance for Section 2 will be to compare your arguments and counter-arguments with the model answer, or let your friends or teachers read it and gather some feedback from them. Fortunately for the mock papers in this book, there are model answers for you to compare your essays against!

Mock Papers

Mock Paper A

Question 1:
"Competitors need to be able to run 200 metres in under 25 seconds to qualify for a tournament. James, Steven and Joe are attempting to qualify. Steven and Joe run faster than James. James' best time over 200 metres is 26.2 seconds."

Which response is **definitely** true?
- A. Only Joe qualifies.
- B. James does not qualify.
- C. Joe and Steven both qualify.
- D. Joe qualifies.
- E. No one qualifies.

Question 2:
You spend £5.60 in total on a sandwich, a packet of crisps and a watermelon. The watermelon cost twice as much as the sandwich, and the sandwich cost twice the price of the crisps.

How much did the watermelon cost?
- A. £1.20
- B. £2.60
- C. £2.80
- D. £3.20
- E. £3.60

Question 3:
Jane, Chloe and Sam are all going by train to a football match. Chloe gets the 2:15pm train. Sam's journey takes twice as long Jane's. Sam catches the 3:00pm train. Jane leaves 20 minutes after Chloe and arrives at 3:25pm.

When will Sam arrive?
- A. 3:50pm
- B. 4:10pm
- C. 4:15pm
- D. 4:30pm
- E. 4:40pm

Question 4:
Michael has eleven sweets. He gives three sweets to Hannah. Hannah now has twice the number of sweets Michael has remaining. How many sweets did Hannah have before the transaction?
- A. 11
- B. 12
- C. 13
- D. 14
- E. 1

Question 5:
"Alex's current weekly take-home pay is £250 per week. Alex is to receive a pay rise of 5% plus an extra £6 per week. The flat rate of income tax will decrease from 14% to 12% at the same time."

What will his new weekly take-home pay be, to the nearest whole pound?
- A. £260
- B. £267
- C. £274
- D. £279
- E. £285

Question 6:
"You have four boxes, each containing two coloured cubes. Box A contains two white cubes, Box B contains two black cubes, and Boxes C and D both contain one white cube and one black cube. You pick a box at random and take out one cube. It is a white cube. You then draw another cube from the same box."

What is the probability that this cube is not white?
- A. ½
- B. ⅓
- C. ⅔
- D. ¼
- E. ¾

Question 7:
"Anderson & Co. hire out heavy plant machinery at a cost of £500 per day. There is a surcharge for heavy usage, at a rate of £10 per minute of usage over 80 minutes. Concordia & Co. charge £600 per day for similar machinery, plus £5 for every minute of usage."

At what duration of usage are the costs the same for both companies?
- A. 100 minutes
- B. 130 minutes
- C. 140 minutes
- D. 170 minutes
- E. 180 minutes

Question 8:
"Simon is discussing with Seth whether or not a candidate is suitable for a job. When pressed for a weakness at interview, the candidate told Simon that he is a slow eater. Simon argues that this will reduce the candidate's productivity, since he will be inclined to take longer lunch breaks."

*Which statement **best** substantiates Simon's argument?*
- A. Slow eaters will take longer to eat lunch
- B. Longer lunch breaks are a distraction
- C. Eating more slowly will reduce the time available to work
- D. Eating slowly is a weakness
- E. People who like food are more likely to eat slowly

Question 9:
Three pieces of music are on repeat in different rooms of a house. One piece of music is three minutes long, one is four minutes long and the final one is 100 seconds long. All pieces of music start playing at exactly the same time.

How long is it until they are next all starting together?
- A. 12 minutes
- B. 15 minutes
- C. 20 minutes
- D. 60 minutes
- E. 300 minutes

Question 10:
A car leaves Salisbury at 8:22am and travels 180 miles to Lincoln, arriving at 12:07pm. Near Warwick, the driver stopped for a 14 minute break.

What was its average speed, whilst travelling, in kilometres per hour? It should be assumed that the conversion from miles to kilometres is 1:1.6.
- A. 51kph
- B. 67kph
- C. 77kph
- D. 82kph
- E. 86kph

Questions **11** and **12** refer to the following data:
Five respondents were asked to estimate the value of three bottles of wine, in pounds sterling.

Respondent	Wine 1	Wine 2	Wine 3
1	13	16	25
2	17	16	23
3	11	17	21
4	13	15	14
5	15	19	29
Actual retail value	8	25	23

Question 11:
What is the mean error margin in the guessing of the value of wine 1?
A. £4.80   B. £5.60   C. £5.80   D. £6.20   E. £6.40

Question 12:
Which respondent guessed most accurately on average?
A. Respondent 1
B. Respondent 2
C. Respondent 3
D. Respondent 4
E. Respondent 5

Questions **13** and **14** refer to the following data
The population of Country A is 40% greater than the population of Country B.
The population of Country C is 30% less than the population of Country D (which has a population 20% greater than Country B).

Question 13:
Given that the population of Country A is 45 million, what is the population of country D?
A. 32.1 million
B. 35.8 million
C. 36.6 million
D. 38.5 million
E. 39.0 million

Question 14:
The population of Country A is still 45 million. If Country B introduced a new health initiative costing $45 per capita, what would be the total cost?
A. $1.34 billion
B. $1.44 billion
C. $1.50 billion
D. $1.56 billion
E. $1.66 billion

Question 15:
A car averages a speed of 30mph over a certain distance and then returns over the same distance at an average speed of 20mph.

What is the average speed for the journey as a whole?
A. 22.5 mph
B. 24 mph
C. 25 mph
D. 26 mph
E. The distance travelled is required to calculate average speed

Question 16:
"All sheep are ruminants and all marsupials are mammals. No sheep are marsupials."

Which of the following must be true?
A. Some ruminants are marsupials.
B. All mammals are marsupials.
C. All sheep are mammals.
D. Some sheep are marsupials.
E. None of the above

Question 17:
The price of toothpaste rises by 80%. This is later reduced by 50% due to competition. Zoe buys two tubes of toothpaste and gets the third free because of a loyalty card.

How much did she have to pay per tube of toothpaste? Express your answer as a percentage of the original price.
A. 16.67%   B. 33%   C. 60%   D. 66.7%   E. 100%

Question 18:
"You can remain fit throughout life if you exercise regularly. Simon does not exercise regularly, so he can never become fit."

Which flawed argument has the same structure as this?
A. "You can speak a foreign language if you learn when young. Simon does not speak a foreign language, so he did not learn when young."
B. "You are never tired if you sleep for 8 hours a night. Simon is tired, therefore he doesn't sleep for 8 hours a night"
C. "You can be a good musician if you practice regularly. Simon does not practice regularly, so he can never be a good musician."
D. "You can be good at sport if you have a natural ability. Simon is good at hockey, therefore he has a natural ability."
E. "Eating five portions of fruit and vegetables daily reduces the risk of heart disease. Simon eats more than this, so he will not develop heart disease."

Question 19:
"Reports of cybercrime are increasing year on year. Last year, police dealt with 250% more cybercrime then the year before. Common complaints relate to inappropriate or defamatory use of social media. To deal with this, many police forces are creating dedicated teams to deal with online offences. A pilot study showed that a dedicated cybercrime team solved cases of cybercrime 40% faster than regular detectives. Therefore the measure will act to suppress the rise in cybercrime."

Which statement best validates the above argument?
A. Solving crimes faster is necessary to keep pace with the increase in crime
B. Solving crimes faster leads to more convictions
C. Solving crimes faster increases police resources to tackle crime
D. Solving crimes faster saves money
E. Solving crimes faster reassures the public of action

Question 20:
"Recently in Kansas, a number of farm animals have been found killed in the fields. The nature of the injuries is mysterious, but consistent with tales of alien activity. Local people talk of a number of UFO sightings, and claim extra terrestrial responsibility. Official investigations into these claims have dismissed them, offering rational explanations for the reported phenomena. However, these official investigations have failed to deal with the point that, even if the UFO sightings can be explained in rational terms, the injuries on the carcasses of the farm animals cannot be. Extra terrestrial beings must therefore be responsible for these attacks."

Which of the following best expresses the main conclusion of this argument?
A. Sightings of UFOs cannot be explained by rational means
B. Recent attacks must have been carried out by extraterrestrial beings
C. The injuries on the carcasses are not due to normal predators
D. UFO sightings are common in Kansas
E. Official investigations were a cover-up

Question 21:
"To make a cake you must prepare the ingredients and then bake it in the oven. You purchase the required ingredients from the shop, however the oven is broken. Therefore you cannot make a cake."

Which of the following arguments has the same structure?
A. To get a good job, you must have a strong CV then impress the recruiter at interview. Your CV was not as good as other applicants, therefore you didn't get the job.
B. To get to Paris, you must either fly or take the Eurostar. There are flight delays due to dense fog, therefore you must take the Eurostar.
C. To borrow a library book, you must go to the library and show your library card. At the library, you realise you have forgotten your library card. Therefore you cannot borrow a book.
D. To clean a bedroom window, you need a ladder and a hosepipe. Since you don't have the right equipment, you cannot clean the window.
E. Bears eat both fruit and fish. The river is frozen, so the bear cannot eat fish.

Question 22:
"Growing vegetables requires patience, skill and experience. Patience and skill without experience is common – but often such people give up prematurely as skill alone is insufficient to grow vegetables, and patience can quickly be exhausted."

Which of the following summarises the main argument?
A. Most people lack the skill needed to grow vegetables
B. Growing vegetables requires experience
C. The most important thing is to get experience
D. Most people grow vegetables for a short time but give up due to a lack of skill
E. Successful vegetable growers need to have several positive traits

Question 23:
"Joseph has a bag of building blocks of various shapes and colours. Some of the cubic ones are black. Some of the black ones are pyramid shaped. All blue ones are cylindrical. There is a green one of each shape. There are some pink shapes."

*Which of the following is definitely **NOT** true?*
A. Joseph has pink cylindrical blocks
B. Joseph doesn't have pink cylindrical blocks
C. Joseph has blue cubic blocks
D. Joseph has a green pyramid
E. Joseph doesn't have a black sphere

Question 24:
Sam notes that the time on a normal analogue clock is 1540hrs. What is the smaller angle between the hands on the clock?

A. 110° B. 120° C. 130° D. 140° E. 150°

Question 25:
A fair 6-faced die has 2 sides painted red. The die is rolled 3 times. What is the probability that at least one red side has been rolled?

A. $8/27$ B. $19/27$ C. $21/27$ D. $24/27$ E. 1

Question 26:
"In a particular furniture warehouse, all chairs have four legs. No tables have five legs, nor do any have three. Beds have not less than four legs, but one bed has eight as they must have a multiple of four legs. Sofas have four or six legs. Wardrobes have an even number of legs, and sideboards have an odd number. No other furniture has legs. Brian picks a piece of furniture out, and it has six legs."

What can be deduced about this piece of furniture?
A. It is a table
B. It could be either a wardrobe or a sideboard
C. It must be either a table or a sofa
D. It must be either a table, a sofa or a wardrobe
E. It could be either a bed, a table or a sofa

Question 27:
Two friends live 42 miles away from each other. They walk at 3mph towards each other. One of them has a pet pigeon which starts to fly at 18mph as soon as the friends set off. The pigeon flies back and forth between the two friends until the friends meet.

How many miles does the pigeon travel in total?
A. 63 B. 84 C. 114 D. 126 E. 252

Question 28:
"Fruit juice contains fibre, vitamins and minerals and can be part of a healthy diet. However, it has been suggested that the high sugar content and acidity negates these benefits by leading to increased rates of dental cavities and hyperactivity in children. If left unchecked, a combination of poor dental hygiene and inappropriate diet can lead to disastrous consequences, including serious infections. On the other hand, many juices contain essential vitamins such as vitamin C which helps the immune system fight infections."

What is the main message from this passage?
A. Children should not drink fruit juice.
B. Fruit juice is harmful to health.
C. Fruit juice is good for health.
D. On balance, we should drink more fruit juice.
E. The overall benefits of fruit juice are unclear.

Question 29:
A complete stationery set includes a pen, a pencil, a geometry set and a pad of paper. Pens cost £1.50, pencils cost 50p, geometry sets cost £3 and paper pads cost £1. Sam, Dave and George each want complete sets, but Mr Browett persuades them to share. Sam and Dave agree to share a paper pad and a geometry set. George must have his own pen, but agrees that he and Sam can share a pencil.

What is the total amount spent?
A. £12.00 B. £13.50 C. £16.50 D. £17.50 E. £18.00

MOCK PAPER A — SECTION ONE

Question 30:
"If the government financially supports the arts, a proportion of each person's taxes will be used to finance museums, galleries and theatres. But some taxpayers have no interest in the arts and never go to theatres or museums. Many of those who enjoy the arts are able to afford to pay for them. Since no one should be forced to subsidise services which they themselves do not use, taxpayers' money should not be used to support the arts."

Which counter-argument provides the strongest rebuke of this principle?
A. If public funding for the arts is withdrawn, only those who are genuinely interested would pay to visit museums
B. The rail network is publically subsidised, although some people do not use trains
C. If people only pay for services they use, then those who can afford private health insurance would not pay towards the NHS
D. Funding museums allows greater preservation of our heritage
E. If something requires subsidy, then people must not genuinely want it

Question 31:
The figure below shows 5 squares made from 12 matches. Which 2 matches need to be moved to make 7 squares?

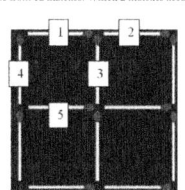

A. 1 and 2
B. 1 and 3
C. 1 and 4
D. 1 and 5
E. Not possible

Question 32:
A cube has six sides of different colours. The red side is opposite to black. The blue side is adjacent to white. The brown side is adjacent to blue. The final side is yellow.

Which colour is opposite brown?
A. Red   B. Black   C. Blue   D. White   E. Yellow

Question 33:
The UK imports 36,000,000 kg of cocoa beans each year. Each g costs the UK 0.3p, from which the supplier takes 20% commission. Of what is left, the local government takes 60% and the distribution company gets 30%.

How much are the cocoa farmers left with per year?
A. £3.68m   B. £6.82m   C. £8.64m   D. £10.8m   E. £11.4m

Questions **34** and **35** refer to the following passage:

- In the year ending June 2013 there were 1,730 fatalities in reported personal injury accidents, a 3% drop from 1,785 in the year ending June 2012. The number of killed or seriously injured (KSI) casualties fell by 5%, to 23,530, and the total number of casualties fell by 7% to 188,540.

- A total of 8,560 car users were reported killed or seriously injured in the year ending June 2013, a fall of 6% from the previous 12-month period.

- KSI casualties for the vulnerable road user groups – pedestrians, pedal cyclists and motorcyclists – showed overall decreases of 7%, 1% and 6% respectively compared with the year ending June 2012.

- The casualty rate per billion vehicle miles decreased for all casualty severities in the year ending June 2013, with falls of 3% for fatalities, 6% for serious injuries and 7% for all casualties. This is the first publication in which the Department has included quarterly casualty rates.

- There were also significant decreases in the number of child casualties (aged 0-15) which fell from 18,166 in the year ending June 2012, to 15,920 in the year ending June 2013, a fall of 12%. The number of child KSIs also fell in the same period by 11% to 2,080. The number of child pedestrian casualties who were killed or seriously injured fell by 8% to 1,440 in the year ending June 2013.

- There were drops in the number of accidents on all road types in the year ending June 2013 relative to the year ending June 2012. The number of fatal or serious accidents fell by 7% on major roads (motorways and A roads) and 4% on minor roads. On roads with speed limits over 40 mph (non-built up) fatal and serious accidents fell by 6% and on roads with speeds limits up to an including 40 mph (built-up) they fell by 5%.

- There were 185,540 casualties from 139,350 accidents in the year ending June 2013 which represents a 6% fall for accidents and a 7% fall for casualties compared with the year ending June 2012.

Question 34:
Regarding the passage, which of these statements can be known to be true?
A. Child casualties are on the rise
B. Annual road deaths in the UK are falling
C. Vulnerable road users are more likely to be injured per vehicle mile than drivers
D. From June 2012 to June 2013, there were 188,540 serious injuries
E. Motorways are safer than built-up roads

Question 35:
"The government is always under pressure to reduce road casualties. For this reason, anti-drink-drive campaigns costing millions of pounds are commonly produced, particularly around Christmas time. To address a one-year increase in drink driving related deaths, a new campaign was introduced. Subsequently, drink-driving casualties fell. The government therefore concluded that the £8m campaign had been a success."

Which of the following most undermines this argument?
A. Fewer people drink-drive these days than 10 years ago
B. Correlation does not imply causation: there is no plausible mechanism for the campaign to provide benefit.
C. The effect is too rapid for this campaign to have changed the public's attitude
D. Regression to the mean can explain this phenomenon: values which were abnormally high one year are likely to settle down the next
E. When spending so much money, benefits are certain. The true test is in running a smaller campaign.

Question 36:
"Some people with a sore throat and a chest infection have the 'flu'."

Which of the following statements is supported?
A. Some people have a chest infection, but do not have the 'flu'
B. Some people with a sore throat and a chest infection do not have the 'flu'
C. Kate has the 'flu'. Therefore she has a sore throat
D. The 'flu' is defined as a sore throat and chest infection together
E. None of the above

Question 37:
Catherine has 6 pairs of red socks, 6 pairs of blue socks and 6 pairs of grey socks in her drawer. Unfortunately, they are not paired together. The light in her room is broken so she cannot see what colour the socks are. She decides to keep taking socks from the drawer until she has a matching pair. What is the minimum number of socks she needs to take from the drawer to guarantee at least one matching pair can be made?
A. 2   B. 3   C. 4   D. 5   E. 6

Question 38:
Luca and Giovanni are waiters. One month, Luca worked 100 hours at normal pay and 20 hours at overtime pay. Giovanni worked 80 hours at normal pay and 60 hours at overtime pay. Neither received any tips. Luca earned €2000; Giovanni earned €2700.

What is the overtime rate of pay?
A. €10 per hour
B. €15 per hour
C. €20 per hour
D. €25 per hour
E. €30 per hour

Question 39:
"Train A leaves Plymouth at 10:00 and travels at 90mph. Train B leaves Manchester at 10:45 and travels at 70mph. The distance between the two cities is 405 miles. Due to a mistake, both trains are travelling on the same track."

Calculate the distance from Plymouth at which the trains will collide.
A. 158 miles   B. 203 miles   C. 228 miles   D. 248 miles   E. 263 miles

Question 40:
"100 pieces of rabbit food will feed one pregnant rabbit and two normal rabbits for a day. 175 pieces of food will feed two pregnant and three normal rabbits for a day. There is no excess food."

Which statement is NOT true?
A. A normal rabbit can be fed for longer than a day with 30 pieces of food.
B. 70 pieces of food are sufficient to feed a pregnant rabbit for a day.
C. A pregnant rabbit needs twice as many pieces per day as a normal rabbit.
D. Two pregnant and four normal rabbits will need 200 pieces of food for a day.
E. Three pregnant and ten normal rabbits will need 450 pieces of food for a day.

Question 41:
"Studies of the brains of London taxi drivers show that training for "the knowledge", a difficult exam requiring knowledge of 20,000 London streets, enlarged a part of the brain believed to be important for spatial and organisational memory. This shows the brain can adapt to training and increase its abilities. Therefore if I wanted to improve my ability to remember names, I should also train my brain with repetitive tasks."

*Which of the following **best** represents the flaw in this argument?*
A. Enlarging of the brain does not necessarily mean it has improved
B. It might not be true to assume name memory and spatial memory use the same part of the brain
C. The brain enlargement would likely have happened anyway even without training
D. We do not know how London taxi drivers prepare for "the knowledge"
E. Practice does not necessarily improve performance on memory tasks

Question 42:
"Michael bought a painting at an auction for £60. After 6 months, he realised the value of the painting had increased, so he sold it for £90. Realising a mistake, he wanted to buy the painting back, which he was able to do for £110. A year later, he then re-sold the painting for £130."

What is the total profit on the painting?
A. £20   B. £30   C. £40   D. £50   E. £60

Question 43:
"Insect pests such as aphids and weevils can be a problem for farmers, as they feed on crops, causing destruction. Thus many farmers spray their crops with pesticides to kill these insects, increasing their crop yield. However, there are also predatory insects such as wasps and beetles that naturally prey on these pests – which are also killed by pesticides. Therefore it would be better to let these natural predators control the pests, rather than by spraying needless chemicals."

Which of the following best describes the flaw in this logic?
A. Many pesticides are expensive, so should not be used unless necessary
B. It fails to consider other problems the pesticides may cause
C. It does not explain why weevils are a problem
D. It fails to assess the effectiveness of natural predators compared to pesticides
E. It does not consider the benefits of using fewer pesticides

Question 44:
A parliament contains 400 members. Last election, there was a majority of 43% of the popular vote to the liberal party. However, as a first-past-the-post system of constituencies was in effect, they gained 298 seats in parliament.

How many excess members did they have, relative to a straight proportional representation system?
A. 72   B. 98   C. 112   D. 126   E. 148

Question 45:
A cube is painted such that no two faces that touch may be the same colour. What is the minimum number of colours required for this?
A. 2   B. 3   C. 4   D. 5   E. 6

Question 46:

MOCK PAPER A — SECTION ONE

4 people need to cross a river, and one of them is fat. They make a stable raft, but find it can only take the weight of either two thin people or the fat person alone. The raft must have someone in it to cross the river in order to propel and steer it.

What is the minimum number of journeys the raft must make across the river to get all 4 people to the other side?
A. 3 B. 5 C. 7 D. 9 E. 11

Question 47:
In a given year, there were four Wednesdays and four Saturdays in December. What day was Christmas Day (25th December)?
A. Monday
B. Tuesday
C. Wednesday
D. Thursday
E. Friday

Question 48:
"Ruddock is West of Langley but East of Dell. Hampton is midway between Langley and Ruddock. Iver is West of Ruddock. Johnstown is not East of Langley."

*Which of the following **cannot** be concluded?*
A. Hampton is East of Iver and Ruddock.
B. Ruddock is West of Langley and East of Iver.
C. Dell is west of Hampton and Langley.
D. Langley is East of Ruddock and East of Hampton
E. Iver is West of Ruddock and West of Dell.

Question 49:
"Zips and buttons are on the opposite side of women's clothing relative to men's. This is because high society always dictated clothing style, and women in high society would historically have had someone to dress them. Therefore the fastenings were positioned for the convenience of the servant and not the wearer. In our age, very few people have a servant to dress them. Therefore buttons and zips on women's clothing should be moved in accordance with the style of men's clothing."

Which of the following statements best describes the principle supporting this argument?
A. The needs of the majority should be of foremost importance
B. It would be more cost effective to make all clothes the same way
C. Traditions are of little value as times change
D. It would be easier for women to fasten clothes if buttons were reversed
E. Style is no longer dictated by high society

Question 50:
"Mark does not drink tea after 9pm as it contains caffeine. Coffee contains more caffeine than tea, therefore Mark does not drink coffee after 9pm either."

Which argument has the same structure?
A. Mark does not like onions. Curry contains onions, therefore Mark does not like curry either.
B. Mark cannot afford a pure wool suit. Pure silk suits are more expensive, so Mark cannot afford a silk suit either.
C. Mark is travelling to London. Brighton is further then London, so Mark is not travelling to Brighton.
D. The bus into town is slow. Therefore Mark will take a taxi there instead.
E. Mark can run faster than Steve. Joe is not as fast as Steve, therefore Mark can run faster than Joe.

END OF SECTION
YOU MUST ANSWER ONLY ONE OF THE FOLLOWING QUESTIONS

Question 1:

"Strive not to be a success, but to be of value"

To what extent is it possible to be "a success", but to have little value?

Question 2:

Is the media a positive or negative influence on scientific understanding?

Question 3:

"Why tell the truth if a lie is better for all concerned?"

In what circumstances can dishonesty be justified?

Question 4:

"Science is a nothing more than a thought process"

What actually is science and how is it of value to us?

END OF TEST

Mock Paper B

Question 1:

MOCK PAPER B   SECTION ONE

"Joseph changes jobs and gets a basic pay cut of 5%, but his tax-free monthly bonus increases from £40 to £90. He also changes tax bracket, so instead of his 10% flat rate, he pays 20% tax on all income over £10,000 pa. Joseph's current weekly take-home pay is £560 per week, exclusive of bonus."

What will his new annual take-home pay be, to the nearest hundred pounds?
A. £25,800   B. £26,500   C. £27,700   D. £29,300   E. £31,300

Question 2:
"Peter books a return flight to Dubai for £725. The flight is refundable, but there is a fee of £45 payable for cancelling. Peter notices as time passes, the remaining tickets on the same plane are becoming cheaper. He decides to cancel his flight, booking a new one for £530 through the same provider. Once again he sees prices have fallen, so he cancels this flight but can only buy a new one for £495."

What is his overall saving, relative to the original price paid?
A. £110   B. £140   C. £150   D. £195   E. £230

Question 3:
"You have three bags, each containing four balls numbered with single digit numbers. Bag A contains even numbers only, Bag B contains odd numbers only, and Bag C contains the numbers 2, 5, 6 and 8. You take a ball from Bag B and put it into Bag C; then you then take a ball from Bag C and put it into Bag A. You draw a ball at random from Bag A."

What is the probability that this ball is an odd number?
A. $1/25$ B. $2/25$ C. $3/25$ D. $4/25$ E. $1/5$

Question 4:
The price of bread rises by 40% due to a poor grain harvest. This is later reduced by 20% due to a government farming subsidy. Dave buys three loaves of bread and gets a fourth free because of a discount in the shop. How much did he pay per loaf of bread? Express your answer as a percentage of the original price.
A. 66%   B. 84%   C. 92%   D. 98%   E. 110%

Question 5:
Sam notes that the time on a normal analogue clock is 2120hrs. What is the smaller angle between the hands on the clock?
A. 130°   B. 140°   C. 150°   D. 160°   E. 170°

Question 6:
Sam needs to measure out exactly 4 litres of water into a tank. He has two pieces of equipment – a bucket that holds 5 litres and a one that holds 3 litres, with no intermediate markings. Is it possible to measure out 4 litres?

If so, how much water is needed in total in order to measure the 4 litres?
A. 4 litres
B. 7 litres
C. 8 litres
D. 10 litres
E. Not possible with this equipment

Question 7:
"A librarian is sorting books into their correct locations. All history books belong to the right of all science books. Science books are divided into five locations: engineering, biology, chemistry, physics and mathematics (in order

~ 35 ~

from right to left). Art books are located between engineering and sport, and sport books between art and history. Literature books are to the right of art books."

What can be said with certainty about the location of literature books?
A. They are located between art and history books
B. They are located to the left of history books
C. They are located between mathematics and art
D. They are located to the right of engineering
E. They are not located to the left of sport

Question 8:
"Many people choose not to buy brand new cars, as buying brand new has significant disadvantages. Most importantly, a car's value drops substantially at the moment it is first driven on the road. Even though a car is virtually unchanged by these first few miles, the potential resale value is significantly reduced. Therefore it is better to buy second hand cars, as their value does not drop so much immediately after purchase."

Which of the following best represents the main conclusion of this passage?
A. There are many equal reasons to avoid buying brand new cars
B. Cars that have driven lots of miles should be avoided
C. The rapid loss of value in new cars makes buying second-hand a wise choice
D. Second hand cars are at least as good as new ones
E. New cars should not be driven to ensure they keep their resale value

Question 9:
James is a wine dealer specialising in French wine. From his original stock of 2,000 bottles in one cellar, he sells 10% to one customer and 20% of the remaining wine to another customer. He makes £11,200 profit from the two transactions combined.

What is the average profit per bottle?
A. £18   B. £20   C. £22   D. £24   E. £26

Question 10:
"Why should we bother exploring deep into the oceans? The programmes are very expensive, and seldom produce any results which benefit normal people. Instead, we should invest resources into supporting people in trouble, rather than wasting money on needless exploration."

Which of the following, if true, would most weaken the above argument?
A. Ocean exploration is less expensive than space exploration, which people are generally happy with
B. Ocean exploration provides fascinating information about bizarre life forms
C. Exploration has led to the discovery of new chemicals which have been used for many new medically useful drugs
D. Exploring into the oceans is safe, given modern submarine technology
E. Money is useful to help people in trouble

Question 11:
"Many good quality pieces of old furniture are considered 'timeless' – they are used and enjoyed by many people today, and this is expected to continue for many generations to come. However, most of this furniture dates back

to previous eras, and modern furniture does not fall under the 'timeless' category of being enjoyed for many years to come."

Which of the following is the main flaw in the argument?
A. There may be many factors which make furniture good
B. There used to be more furniture makers than today
C. No evidence is given to tell us old furniture is better than new
D. Old furniture is desirable for other reasons than its quality
E. We cannot yet tell whether new furniture will become 'timeless'

Question 12:
"Red wine is thought to be much healthier than beer because it contains many antioxidants, which have been shown to be beneficial to health. Many red wines are produced in Southern France and Italy, therefore it is no surprise that residents there have a greater life expectancy than in the UK and Germany, which are predominantly beer producing countries."

Which of the following is an assumption of the above argument?
A. Italian people drink red wine
B. Antioxidants are beneficial for health
C. British people prefer beer to red wine
D. Beer is not produced in Italy
E. Italian life expectancy is greater than in the UK

Question 13:
Hannah, Jane and Tom are travelling to London to see a musical. Hannah catches the train at 1430. Jane leaves at the same time as Hannah, but catches a bus which takes 40% longer then Hannah's train. Tom also takes a train, and the journey time is 10 minutes less then Hannah's journey, but he leaves 45 minutes after Jane leaves. He arrives in London at 1620.

At what time will Jane arrive in London?
A. 1545   B. 1600   C. 1615   D. 1700   E. 1715

Question 14:
At a show, there are two different ticket prices for different seats. The cost is £10 for a standard seat, and £16 for a premium view seat. The total revenue from a show is £6,600, and the total attendance was 600.

How many premium view seats were purchased?
A. 60   B. 100   C. 140   D. 180   E. 240

Question 15:
The moon orbits the Earth once every 28 days. Between 20th January and 23rd April inclusive, how many degrees has the Moon turned through? This is not a leap year.
A. 1010   B. 1100   C. 1210   D. 1500   E. 1620

Question 16:
Drama academies are special schools that students can go to in order to learn performing arts. These schools are only available to the most skilled young performers, and aim to give students the best training in the arts, whilst still covering mainstream academic subjects. However, many parents are reluctant for their children to attend such academies, as they feel the academic teaching will be worse than a standard school.

Which of the following, if true, would most weaken the above argument?
A. Most top actors attended a drama academy as children
B. There is as much time dedicated to academic work in drama academies as there is in normal schools
C. The academic work comprises a greater proportion of the study time than drama related activities
D. Most children are keen to attend a drama academy if given the opportunity
E. 80% of students at drama academies attain higher than average GCSE scores

Question 17:
Anil and Suresh both leave point A at the same time. Anil travels 5km East then 10km North. Anil then travels a further 1km North before heading 3km West. Suresh travels East for 2km less than Anil's total journey distance. He then heads 13km North, before pausing and travelling back 2km South. How far, as the crow flies, are the two men now apart?
A. 11km   B. 12km   C. 13km   D. 15km   E. 17km

Question 18:
Building foundations are covered by 14cm of concrete. A builder thinks this is too thick, and grinds down the concrete by an amount three times the thickness of the concrete which he eventually leaves.

What is the remaining thickness of concrete?
A. 1.5cm   B. 2.0cm   C. 2.5cm   D. 3.0cm   E. 3.5cm

Question 19:
Chris leaves his house to go and visit Laura, who lives 3 miles away. He leaves at 1730 and walks at 4mph towards Laura's house, stopping once for a 5-minute chat to a friend. Meanwhile Sarah also wants to visit Laura. She sets off from her house 6 miles away at 1810, driving in her car and averaging a speed of 24mph.

Who reaches the house first and with how long do they wait for the other person?
A. Chris, and waits 5 mins for Sarah
B. Chris, and waits 10 mins for Sarah
C. Sarah, and waits 5 mins for Chris
D. Sarah, and waits 10 mins for Chris
E. They both arrive at the same time

Question 20:
"Illegal film and music downloads have increased greatly in recent years. This causes significant harm to the relevant industries. Many people justify this to themselves by telling themselves that they are only diverting money away from wealthy and successful singers and actors, who do not need any more money anyway. But in reality, illegal downloads are deeply harming the music industry, making many studio workers redundant and making it difficult for less famous performers to make a living."

Which of the following best summarises the conclusion of this argument?
A. Unemployment is a problem in the music industry
B. Taking profits away from successful musicians does more harm than good
C. Studio workers are most affected by illegal downloads
D. Illegal downloads cause more harm than people often think
E. Buying music legally helps keep the music industry productive

Question 21:
"40,000 litres of water will extinguish two typical house fires. 70,000 litres of water will extinguish two house fires and three garden fires. There is no surplus water"

*Which statement is **NOT** true?*
A. A garden fire can be extinguished with 12,000 litres, with water to spare.
B. 20,000 litres is sufficient to extinguish a normal house fire.
C. A garden fire requires only half as much water to extinguish as a house fire.
D. Two house and four garden fires will need 80,000 litres to extinguish.
E. Three house and ten garden fires will need 140,000 litres to extinguish.

Question 22:
A car travels at $20ms^{-1}$ for 30 seconds. It then accelerates at a constant rate of $2ms^{-2}$ for 5 seconds, then proceeds at the new speed for 20 seconds before braking with constant deceleration of $3ms^{-2}$ to a stop.

What distance is covered in total?
A. 1325m   B. 1350m   C. 1375m   D. 1425m   E. 1475m

Question 23:
"Plans are in place to install antennas underground, so that users of underground trains will be able to pick up mobile reception. There are, as usual, winners and losers from this policy. Supporters of the policy argue that it will lead to an increase in workforce productivity and will increase convenience in day-to-day life. Critics respond by saying that it will lead to an annoying environment whilst travelling, it will more easily facilitate a terrorist threat and it will decrease levels of sociability. The latter camp seems to have the greater support and so a re-consideration of the policy is urged."

*Which of the following **best** summarises the conclusion of this passage?*
A. The disadvantages of installing underground antennas outweigh the benefits
B. The cost of the scheme is likely to be prohibitive
C. The policy must be dropped, since the majority do not want it
D. More people don't want this scheme than do want it
E. A detailed consultation process should take place

Question 24:
"Ecosystems in the oceans are changing. Recently, restrictions on fishing have been imposed to tackle the decline in fish populations. As a result, farm fishing and the price of fish have increased whilst the seas recover. It is hoped that these changes will lead to a brighter future for all."

*Which of the following are **TWO** assumptions of this argument?*
*PLEASE MARK **TWO** RESPONSES*
A. People will still buy farmed fish at a higher price
B. The population of wild fish can recover
C. Fishermen will benefit from working on this scheme
D. Ecosystems have been altered as a result of climate change
E. Heavy sea fishing is to blame for the changes in the ecosystem

Question 25:
Brian is tossing a coin. He tosses the coin 5 times. What is the probability of tossing exactly 2 heads?
A. $^1/_{16}$ B. $^5/_{32}$ C. $^4/_{16}$ D. $^5/_{16}$ E. $^7/_{16}$

Question 26:
The amount of a cleaning powder to be added to a bucket of water is determined by the volume of water, such that exactly 40g is added to each litre. A bucket contains 5 litres of water, and is required to have cleaning powder added. However, the markings on the bucket are only accurate to the nearest 2%. Calculate the difference between the maximum and minimum amounts of cleaning powder which might be required to be added to make up the solution correctly.

A. 4g   B. 6g   C. 8g   D. 12g   E. 20g

Question 27:
International telephone calls are charged at a rate per minute. For a call between two European countries, the rate is 22p per minute off-peak and 32p per minute at peak hours, rounded up to the nearest whole minute. In addition, there is a connection fee of 18p for every call.

What is the cost of an off-peak call from France to Germany, lasting 1.4 hours?

A. £18.48   B. £18.66   C. £26.88   D. £27.06   E. £30.98

Question 28:
"UV radiation is harmful to the skin, and can lead to the development of skin cancers. Despite this, many people sunbathe and use tanning salons, exposing themselves to dangerous radiation. If people took more sensible decisions about their health, many serious diseases, such as skin cancers, could be avoided."

What is the main conclusion of this passage?
A. UV radiation is harmful to the skin
B. Many people like to get tanned, despite the risks
C. People do not always consider the health risks of choices they make
D. Skin cancer is a serious disease
E. Sunbathing is risky, and people should avoid it

Question 29:
"Today it is raining. My umbrella is broken; therefore I will wear an anorak."

Which of the following arguments follows the same structure?
A. I want to go to London. The train is delayed, therefore I will be late
B. The end-of-year exam is difficult. I want to do well, so will study hard.
C. My clothes are wet. The tumble dryer is in use, therefore I will pin the clothes on the washing line.
D. The piano is old. Because it is out of tune, I will not play it.
E. Bleach is an effective cleaning product. However there is also soap, which is better for washing hands.

Question 30:
Jim washes windows for pocket money. Washing a window takes two minutes. Between one house and the next, it takes Jim 15 minutes to pack up, walk to the next house and get ready to start washing again. Each resident pays Jim £3 per house, regardless of how many windows the house has. In one day, Jim washes 8 houses, with an average of 11 windows per house.

What is his equivalent hourly pay rate?

A. £4.38   B. £4.86   C. £5.12   D. £5.62   E. £6.12

Question 31:
"Bottled water is becomingly increasingly popular, but it is hard to see why. Bottled water costs many hundreds of times more than a virtually identical product from the tap, and bears a significant environmental cost of transportation. Those who argue in favour of bottled water may point out that the flavour is slightly better – but would you pay 300 times the price for a car with just a few added features?"

Which of the following, if true, would most weaken the above argument?
A. Bottled water has many health benefits in addition to tasting nicer
B. Bottled water does not taste any different to tap water
C. The cost of transportation is only a fraction of the costs associated with bottling and selling water
D. Some people do buy very expensive cars
E. Buying bottled water supports a big industry, providing many jobs to people

Question 32:
"There are no marathon runners that aren't skinny, nor no cyclists that aren't marathon runners."

*Which of the following **must** be true?*
A. Cyclists do not run marathons
B. Cyclists are all skinny
C. Any skinny person is also a cyclist
D. Marathon runners must all be cyclists
E. All of the above

Question 33:
"Langham is East of Hadleigh but West of Frampton. Oakton is midway between Langham and Stour. Frampton is West of Stour. Manley is not East of Langham."

*Which of the following **cannot** be concluded?*
A. Oakton is East of Langham and Hadleigh.
B. Frampton is West of Stour and East of Manley.
C. Stour is East of Hadleigh and Langham.
D. Oakton is East of Langham and West of Frampton
E. Manley is West of Oakton and West of Frampton.

Question 34:
A pot of paint gives sufficient paint to cover 12m² of wall area. The inner surface of a planetarium must be painted. The planetarium consists of a hemispheric dome of internal diameter 14 metres. How many pots of paint are required to give the dome two full coats of paint? [Assume $\pi=3$]; [Surface area of sphere = $4\pi r^2$].
A. 25   B. 36   C. 49   D. 64   E. 98

Question 35:
A planetarium has just been painted as in **34**, above. Assuming each pot of paint is 2 litres, and that the solid component of the paint is 40%, calculate the percentage decrease in the volume of the planetarium, due to the painting.
[Assume $\pi=3$]; [Volume of sphere = $4/3 \pi r^3$]; [1 litre = $0.001m^3$].
A. 0.0029%   B. 0.0057%   C. 0.029%   D. 0.057%   E. 2.86%

Question 36:
A sweet wrapping machine takes 400ms to wrap a sweet. How many sweets can it wrap in 2 hours?
A. 3,000   B. 7,000   C. 9,000   D. 14,000   E. 18,000

Question 37:
John's brother is 6 years younger than him. In 8 years time, the sum of their ages will be 52. How old is John now?
A. 15　　　　　B. 18　　　　　C. 21　　　　　D. 24　　　　　E. 26

Question 38:
"In the UK there is currently a housing shortage. This has increased the prices of many homes, making it harder for first time buyers to get on the property ladder than ever before. Some people attribute this to the increased divorce rate and the breakdown of the family unit as with more people living on their own, the number of occupied houses is increased. To solve the issue, more new housing needs to be built."

Which of the following best expresses the main conclusion of this passage?
A. There are not enough houses in the UK
B. Building more new housing will reduce house prices
C. The increased divorce rate is the main reason for housing shortages
D. The current rate of house building is low
E. Divorcees often live in the type of house first time buyers desire

Question 39:
A train travels from Crabtree to Eppingsworth. There are four stations in between at which the train stops. The time taken to travel between each of these stations decreases by one-fifth for each leg of the journey. Travelling from Station 3 to Station 4 takes 16 minutes.

What is the total journey time?
A. 80 minutes
B. 89 minutes
C. 105 minutes
D. 125 minutes
E. 183 minutes

Question 40:
HS2 is a proposal for a new high speed rail link between London and Birmingham. Supporters say it is required to increase capacity on the congested existing line, and will bring economic benefits through decreased journey times. However many of those living near the proposed development are concerned. They fear the construction will bring a lot of noise and be a visual scar on the landscape, in the same way that the refurbishments of the old line did.

Which of the following, if true, would most appease the residents?
A. The benefits of fast rail links would be felt by all
B. If the line was moved elsewhere, other people would be similarly affected
C. Journey times will be reduced significantly
D. Modern trains are quieter than old ones
E. Since the last works, construction equipment has become quieter and there are more stringent regulations against noise and mess

Question 41:
Van hire from Tony's costs £23 per day. However longer term van hire is cheaper from Adam's, at a rate of £18 per day, but with an initial rental fee of £65.

How many days would you have to hire for to make a saving by hiring from Adam's rather than Tony's?
A. 9 days　　　　B. 11 days　　　　C. 13 days　　　　D. 14 days　　　　E. 17 days

Question 42:
"Antibiotic resistance is on the increase. As a result, many of the antibiotics in our vast armoury are becoming ineffective against common infections. Probably the most significant reason for this is the use of antibiotics in farming, as this exposes bacteria to antibiotics for no good reason, giving the opportunity for resistance to develop. If this worrying trend continues, we might, in 30 years' time, be back in the Victorian situation, where people die from skin or chest infections we consider mild today."

Which of the following best represents the overall conclusion of the passage?
A. Antibiotic resistance is a serious issue
B. Antibiotics use in farming is essential
C. The use of antibiotics in farming could cause us serious harm
D. Victorians used to die from diseases we can treat today
E. Antibiotics can treat skin infections

Question 43:
In a different language, the word for "AMBULANCE" is conveniently written so that it reads the same when viewed directly, and through the rear view mirror of a car. Which of the following is a viable translation for "AMBULANCE"?
 1) TANANAT
 2) AMATAMA
 3) MARAM
 4) ITRARTI
 5) SOOVOOS

A. 1 only   B. 1 and 3   C. 2 only   D. 2 and 5   E. 1, 2 and 5

Question 44:
Sam and Pete are vegetable merchants. Sam buys the vegetables and Pete sells them on. Their trading across 3 weeks is summarised here. In the first week, Sam buys £7,500 of vegetables and Pete sells these for £10,500. In week 2, Sam buys 60% more in value, and Pete sells these with a 30% profit margin. In week 3 the total sale value is £2,000 less than in week 2 – but the profit margin is a healthy 60%.

What value of vegetables did Sam buy in week 3?
A. £7,750   B. £8,500   C. £8,900   D. £9,200   E. £9,750

Question 45:

The following net folds to make a cuboctahedron. This shape is folded up and made into a die. The area of each triangle is $3\sqrt{27}$cm². Assuming the probability of landing on a side is directly proportional to the surface area of that side, calculate the probability of landing on side **X**, expressed in its simplest form.

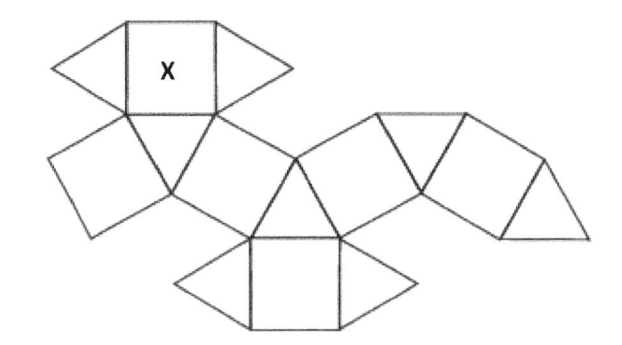

A. $\dfrac{4}{24+\sqrt{27}}$ B. $\dfrac{36}{216+24\sqrt{27}}$ C. $\dfrac{9}{54+\sqrt{27}}$ D. $\dfrac{1}{10}$ E. $\dfrac{24\sqrt{27}}{14^2}$

Question 46:

"Many people commute to work – that is to say they have a repetitive journey between their home and work. In the UK, over 50% of commuters complain about the duration of their commute. The median length of a commute in the UK is 40 minutes, which isn't that long if you think about it. Many of us waste more time than that every day for other reasons, and on the train in particular the time can be used productively."

Which of the following can be reliably concluded from this passage?
A. British people like to complain, even when there are no real problems
B. At least 50% of commuters have a commute which is not longer than 40 minutes
C. People with a commute over 40 minutes are likely to complain about it
D. 50% of people feel the distance to work is too great
E. More than half the population complain about their commute

Question 47:

In a game of bagatelle, Josie scores zero once in every three turns. She takes three turns. What is the probability she scores zero at least once?

A. $^{13}/_{27}$ B. $^{15}/_{27}$ C. $^{19}/_{27}$ D. $^{24}/_{27}$ E. 1

Question 48:
Study this graph, showing the filling of a type of container. Which type of container is being filled?

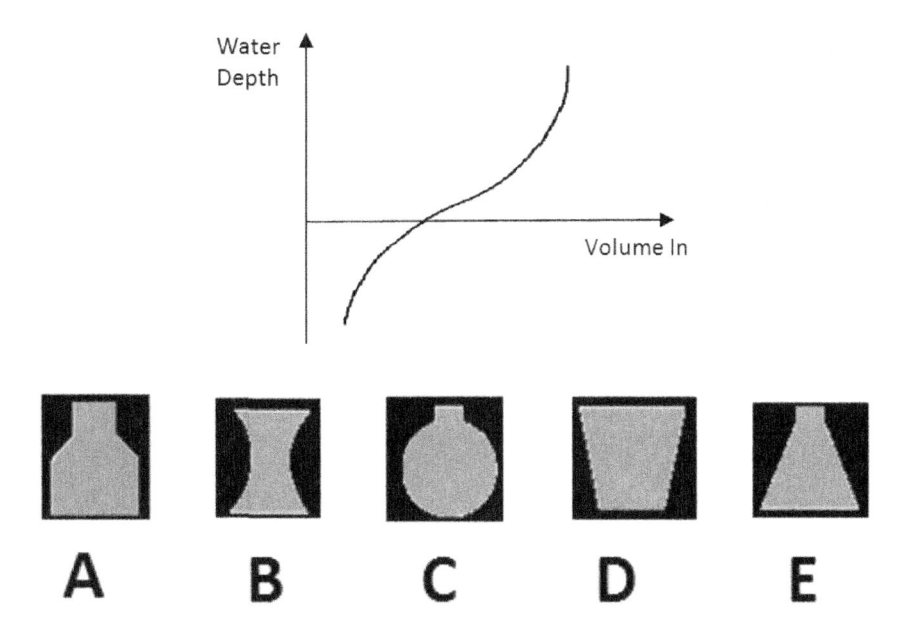

Question 49:
"There has been an increase in the number of listed buildings in the UK. Listed buildings are registered buildings of aesthetic or historical interest, which have restrictions placed on making changes to the building. Therefore it is bad to own a listed building, as it restricts the freedom with which you can modify your house as you wish.

Which of the following is the main flaw in the above argument?
A. Listed buildings are not used as homes
B. It assumes people want to modify their houses
C. It does not take account of the benefits of owning a listed building, such as funding for restorations or higher market value
D. There is no evidence that modifying a listed building is any more difficult
E. Buildings of historical interest are unlikely to need modifications

Question 50:
Scott believes the number of Ford Escorts on the roads decreases by 25% every year. In 2005, there were 36,000 hard top Escorts remaining.

How many does Scott expect will remain in 2015?
A. 1,450   B. 2,030   C. 2,420   D. 2,700   E. 3,600

END OF SECTION

MOCK PAPER B — SECTION TWO

YOU MUST ANSWER ONLY ONE OF THE FOLLOWING QUESTIONS

Question 1:
Design an experiment to deduce the sensitivity of a snake's hearing. Explain everything you would do, and your rationale for doing so.

Question 2:
"The eternal mystery of this world is its comprehensibility"
        To what extent is the world comprehensible?

Question 3:
"The greatest obstacle to learning is education"
        Argue for or against this statement.

Question 4:
Does a vacuum really exist?

END OF TEST

Mock Paper C

Question 1:
"Every year, there are tens of thousands of motor crashes, causing a serious number of deaths, representing the leading cause of death in the UK that is not a disease. However, in spite of this horrendous statistic, there are still thousands of uninsured drivers. The government is under a moral obligation to clamp down on uninsured drivers, to reduce the incidence of such crashes. That they have not acted is arguably the most outrageous failing of the present government."

Which of the following is the best statement of a flaw in this passage?

A. It has made unsupported claims that the government's failure to act is morally outrageous.
B. It has not provided any evidence to support its claims that motor crashes are the leading cause of death in the UK outside of diseases.
C. Even if motor crashes were prevented, it would not save lives of people who die from other causes.
D. It has implied that lack of insurance is related to the incidence of motor crashes.
E. It has fabricated an obligation on the government's part to intervene and reduce the numbers of uninsured drivers.

Question 2:
"Several years ago the Brazilian government held a referendum of the populace, to decide whether they should enact a law banning the ownership of guns. The Brazilian people voted strongly against this proposal. When asked why this had happened, one commentator said he believed the reason was that 90% of criminals who use guns to commit crimes buy their weapons on the black market, illegally. Thus, if Brazil were to ban the legal sale of guns, this would remove the ability of law-abiding citizens to purchase protection, whilst doing little to remove weapons from the hands of criminals.
Some commentators have pointed to this reasoning, and claimed that the UK should also legalise guns, to allow citizens to protect themselves. However, in the UK the black market for weapons is not as widespread as in Brazil. Thus, most people in the UK have little reason to fear gun attacks, and legalising the sale of guns would simply make it much easier for criminals to acquire weapons. Therefore, the situation in Brazil is not applicable to the UK, and legalising gun ownership in the UK would be a bad move."

Which of the following is best supported by this passage?

A. The UK should not legalise guns.
B. The UK should legalise guns.
C. Brazil should ban the ownership of guns.
D. Brazil should not ban the ownership of guns.
E. None of these statements are supported by the passage.

Question 3:
Hannah is buying tiles for her new bathroom. She wants to use the same tiles on the floor and all 4 walls, and for all the walls to be completely tiled apart from the door. The bathroom is 2.4 metres high, 2 metres wide and 2 metres long, and the door is 2 metres high, 80cm wide and at the end of one of the 4 identical walls. The tiles she wants to use are 40cm x 40cm.

How many of these tiles does she need to tile the whole bathroom?
A. 110   B. 120   C. 135   D. 145   E. 150

Question 4:
Jane and Trevor are both travelling south, from London to York. Jane is driving, whilst Trevor is travelling by train. The speed limit on the roads between York and London is 70mph, whilst the train travels at 90mph. Thus, we should expect that Trevor will arrive first.

Which of the following would weaken this passage's conclusion?
A. The train takes a more direct route, whilst the road from York to London goes through several major cities and zig-zags somewhat on its way down the country.
B. Trevor left before Jane.
C. Jane is a conscientious driver, who never exceeds the speed limit.
D. Trevor's train makes a lot of stops on the way, and spends several minutes at each stop waiting for new passengers to board.
E. Meanwhile, Raheem is making the same journey by plane, and will arrive before either Trevor or Jane.

Question 5:
A recipe for 20 cupcakes requires 200g of butter, 200g of sugar, 200g of flour and 4 eggs. Jeremy has two 250g packs of butter, a bag of 600g of sugar, a kilogram bag of flour and a pack of 12 eggs. How many cupcakes can he make and how many eggs does he have left over?
A. 50, 2   B. 50, 3   C. 60, 0   D. 60, 2   E. 60, 3

Question 6:
ABC taxis charges a rate of 15p per minute, plus £4. XYZ taxis charges a rate of £4, plus 30p per mile. I live 6 miles from the station. What would the taxi's average speed have to be on my journey home from the station for the two taxi firms to charge exactly the same fare?
A. 20mph   B. 30mph   C. 40mph   D. 50mph   E. 60mph

Question 7:
"King Arthur has been issued a challenge by Mordac, his nephew who rules the adjacent Kingdom. Mordac has challenged King Arthur to select a knight to complete a series of challenging obstacles, battling a number of dark creatures along the way, in a test known as the Adzol. The King's squire reports that there are tales told by the elders of the court meaning that only a knight with tremendous courage will succeed in Adzol, and all others will fail. He therefore suggests that Arthur should select Lancelot, the most courageous of all Arthur's Knights. The squire argues that due to what the Elders have said, Lancelot will succeed in the task, but all others will fail."

Which of the following is NOT an assumption in the squire's reasoning?
A. Lancelot has sufficient courage to succeed in the Adzol.
B. No other knights in Arthur's command also have tremendous courage, so will all fail Adzol.
C. Great courage is required to be successful in the Adzol.
D. The tales told by the elders of the court are correct.
E. None of the above – they are all assumptions.

Question 8:
A historian is examining a recently excavated hall beneath a medieval castle. She finds that there are a series of arch-shaped gaps along one length of the wall, surrounded by a different pattern of bricks to that seen elsewhere in the walls. These are found to represent where windows were once located, looking out onto one side of the castle. However, the site is now underground. Underground halls in castles never contain windows, so the historian reasons that this hall must once have been located above the ground. Therefore, the ground level must have changed since this castle was built.

Which of the following represents the main conclusion of this passage?
A. Windows are never found in underground halls.
B. Arch-shaped gaps always indicate that windows were once present.
C. It is unexpected for windows to be found in halls in castles.
D. The hall was once located above ground.
E. The ground level must have changed since this hall was built.

Question 9:
Adam's grandmother has sent him to the shop to buy bread rolls. Usually, bread rolls are 30p for a pack of 6 and so his grandmother has given him the exact amount to buy a certain number of bread rolls. However, today there is a special offer whereby if you buy 3 or more packs of rolls, the price per roll is reduced by 1p. He can now buy 1 more pack than before and get no change.
How many bread rolls was he originally supposed to buying?
A. 4 B. 5 C. 6 D. 24 E. 30

Question 10:
"The England men's cricket team have recently been knocked out of the world cup after a very poor performance that saw them eliminated at the group stage, managing only 1 win and losing against teams well below them in the rankings. The board of English cricket is sitting down to discuss why the team's performance was so poor, and what can be done to ensure that future world cups have a more positive outcome. The chairman of the board says that the current crop of players is not good enough, and that the team's performance should improve soon, as more able players come through the ranks in the county teams, so no action is needed.
However, the sporting director takes a different view, saying that England have not gone further than the group stage of any cricket world cup for the last 25 years, during which time numerous players have come and gone from the team. The sporting director argues that this long period of poor performance indicates that there is a problem with English cricket, meaning that not enough talented players are being produced in the country. He argues that therefore, steps should be taken to reform English cricket to actively foster the development of more talented players."

Which of the following, if true, would most strengthen the sporting director's argument?
A. The English cricket team is currently regarded as one of the best in the world, with some of the most talented players.
B. England have been steadily falling lower in the world cricket rankings for the last 25 years, due to poor performances across the board in various cricket competitions.
C. A skilled batsman, who was ranked as the 4th best player in the world, has recently retired from the England team. Now, there are no English cricket players in the top 10 of the world cricket player rankings, which is the first time this has happened in over 70 years.
D. Despite not performing well in world cups, England have performed well in other cricket competitions over the last 20 years.
E. Cricket was invented in England, so everybody expects that England should have a lot of good players in their team.

Question 11:
Karl is making cupcakes for a wedding. It takes him 25 minutes to prepare each batch of cakes. Only 12 can go in the oven at a time and each batch takes 20 minutes in the oven. If he needs to make 100 cupcakes by 4pm, at what time should he start?

A. 11:55am B. 12:20pm C. 12:40pm D. 13:20pm E. 14:00pm

Question 12:

	Boys Absenteeism	Girls Absenteeism	Pupils on roll	Average
Hazelwood Grammar	7%	Boys' school	300	7%
Heather Park Academy	5%	6%	1000	5.60%
Holland Wood Comprehensive	5%	6%	500	5.60%
Hurlington Academy	Girls' school		200	
Average		7%		

Some of the information is missing from the above table. What is the rate of girls' absenteeism at Hurlington Academy?

A. 6.5% B. 7% C. 9% D. 11.5% E. 13%

Question 13:
"Up until the 20th century, all watches were made by hand, by watchmakers. Watchmaking is considered one of the most difficult and delicate of manufacturing skills, requiring immense patience, meticulous attention to detail and an extremely steady hand. However, due to the advent of more accurate technology, most watches are now produced by machines, and only a minority are made by hand, for specialist collectors. Thus, some watchmakers now work for the watch industry, and only perform *repairs* on watches that are initially produced by machines."

Which of the following cannot be reliably concluded from this passage?

A. Most watches are now produced by machines, not by hand.
B. Watchmaking is considered one of the most difficult of manufacturing skills
C. Most watchmakers now work for the watch industry, performing repairs on watches rather than producing new ones.
D. The advent of more accurate technology caused the situation today, where most watches are made by machines.
E. Some watches are now made by hand for specialist collectors.

Question 14:
"Many vegetarians claim that they do not eat meat, poultry or fish because it is unethical to kill a sentient being. Most agree that this argument is logical. However, some Pescatarians have also used this argument, that they do not eat meat because they do not believe in killing sentient beings, but they are happy to eat fish. This argument is clearly illogical. There is powerful evidence that fish fulfil just as much of the criteria for being sentient as do most commonly eaten animals, such as chicken or pigs, but that all these animals lack certain criteria for being "sentient" that humans possess. Thus, Pescatarians should either accept the killing of beings less sentient than humans, and thus be happy to eat meat and poultry, or they should not accept the killing of any partially sentient beings, and thus not be happy to eat fish."

Which of the following best illustrates the main conclusion of this passage.
A. The argument that it is unethical to eat meat due to not wishing to kill sentient beings but eating fish is acceptable, is illogical.
B. Pescatarians cannot use logic.
C. Fish are just as sentient as chicken and pigs, and all these beings are less sentient than humans.
D. It is not unethical to eat meat, poultry or fish.
E. It is unethical to eat all forms of meat, including fish and poultry.

Question 15:
"Recent research into cultural attitudes in Britain has revealed a striking hypocrisy. When asked whether foreign people travelling to Britain on holiday should learn some English, 60% of respondents answered yes. However, when asked if they would attempt to learn some of the language before travelling to a country which did not speak English, only 15% of the respondents answered yes. This is a shocking double-standard on the part of the British public, and is symptomatic of a deeper underlying issue that British people feel themselves superior to other cultures."

Which of the following can be reliably concluded from this passage?
A. 60% of people in Britain think that foreign people travelling to Britain for a holiday should learn English, but would not learn the language themselves when going on holiday to a country which did not speak English.
B. The British public do not feel that it is important to learn some of the language before travelling to a country which does not speak English.
C. There are numerous issues of racism amongst the British public, stemming from the fact they feel themselves superior to other cultures.
D. Less than 10% of the British public would attempt to learn some of the language before travelling to a country which did not speak English.
E. Some in Britain think that foreign people travelling to Britain for a holiday should learn English, but would not learn the language themselves when going on holiday to a country which did not speak English.

Question 16:
Harriet is a headmistress and she is making 400 information packs for the sixth form open evening. Each information pack needs to have 2 double sided sheets of A4 of general information about the school. She also needs to produce 50 A5 single sided sheets about each of the 30 A Level courses on offer. Single sided A5 costs £0.01 per sheet to print. Double sided costs twice as much as single sided. A4 printing costs 1.5 times as much as A5.
How much does she spend altogether on the printing?
A. £270　　　B. £310　　　C. £350　　　D. £390　　　E. £430

Question 17:

"Kirkleatham Town football club are currently leading the league. One week they play a crucial match against Redcar Rovers, who are second placed. The points tally of the teams in the table means that if Kirkleatham Town win this game, they will win the league. Before the game, the manager of Kirkleatham Town says that Redcar Rovers are a tough opponent, and that if his team do not play with desire and commitment they will not win the game. After the game, the manager is asked for comment on the game, and says he was pleased that his team played with so much desire, and showed high levels of commitment. Therefore, Kirkleatham will win the league."

Which of the following best illustrates a flaw in this passage?
A. It has assumed that Kirkleatham will not win the game if they do not play with desire and commitment.
B. It has assumed that if Kirkleatham play with desire and commitment, they will win the game.
C. It has assumed that Kirkleatham played with desire and commitment.
D. It has assumed that Redcar Rovers are a tough opponent, and that Kirkleatham will not be able to easily win the game.
E. It has assumed that if Kirkleatham win the match against Redcar Rovers, they will win the league.

Question 18:
Two councillors are considering planning proposals for a new housing estate, to be built on the edge of Bluedown Village. Councillor Johnson argues for a proposal to be built upon brownfield land, land which has previously been built on, rather than greenbelt land, which has not previously been built on. He argues that this will both lower the cost of building the estate, as the land would already have some underlying infrastructure and would not need as much preparation, and will ensure a minimal impact on wildlife around the area.

Which of the following would most weaken the councillor's argument?
A. Brownfield land is often not as appealing as greenbelt land visually, and it is likely that houses built on brownfield land will not sell for as high a price as houses built on greenbelt land.
B. An area of brownfield land on the edge of the village, originally built as an outdoor leisure complex, has since become run down, and ironically is now a haven for various types of rare newts, lizards and birds.
C. Much of the brownfield land around the edge of the village has undergone substantial underground development, with a good system of electricity cables, gas pipes and plumbing in place.
D. The village is surrounded by several greenbelt areas designated as areas of outstanding natural beauty, supporting an abundance of wildlife.
E. The village mayor, who has ultimate control over the planning proposal, agrees with councillor Johnson's argument. Thus, it is likely his recommendations will be followed

Question 19:

MOCK PAPER C SECTION ONE

John is driving down the A1(M), southbound, having joined at Darlington. If he leaves the motorway at Junction 48, he will arrive at Wetherby Service station. If he goes to Wetherby service station, he will purchase a new cuddly soft toy for his son David. Thus, if John leaves the motorway at Junction 48, he will purchase a new cuddly soft toy.

Which of the following most closely follows the reasoning in this passage?
A. Lucy is travelling from London to Cambridge. She will have to travel on the M25 unless she travels by train. However, Lucy opts to travel by coach, so she will therefore travel on the M25.
B. James is a cricket player working on his batting technique. He finds that whenever he is delivered a ball that does not have any spin on it, he manages to score a run. Whenever he scores a run, his coach shows him a replay, so that he can have a second look at the technique he used. Thus, if James is delivered a ball without spin, he will be shown a replay by his coach.
C. A truck is delivering a cargo of fresh fruit to a supermarket. The driver knows he must deliver the shipment before 8am. If he takes the motorway, he will make the journey in 1 hour less time than if he takes the country lane. Thus, if he takes the motorway, he will be able to leave an hour later.
D. Cleveland police has been informed by the government that its funding is to be cut, and is discussing how best to deal with these cuts. The chief constable knows that if they do not maintain funding of their 999 response vehicles, then the public will be endangered. If this funding is to be maintained, then there will have to be cuts in police patrols. Thus, if the public are not to be endangered, there will have to be cuts in police patrols.
E. If a car's fuel injectors are not properly maintained, then it will be unable to deliver fuel to the engine. If it is unable to deliver fuel to the engine, it will not be able to run, as the engine cannot provide power without fuel. Thus, if the fuel injectors are not properly maintained, the car will not be able to run.

Question 20:
The graph below shows the voting intentions of some constituents interviewed by a polling group, prior to an upcoming election.

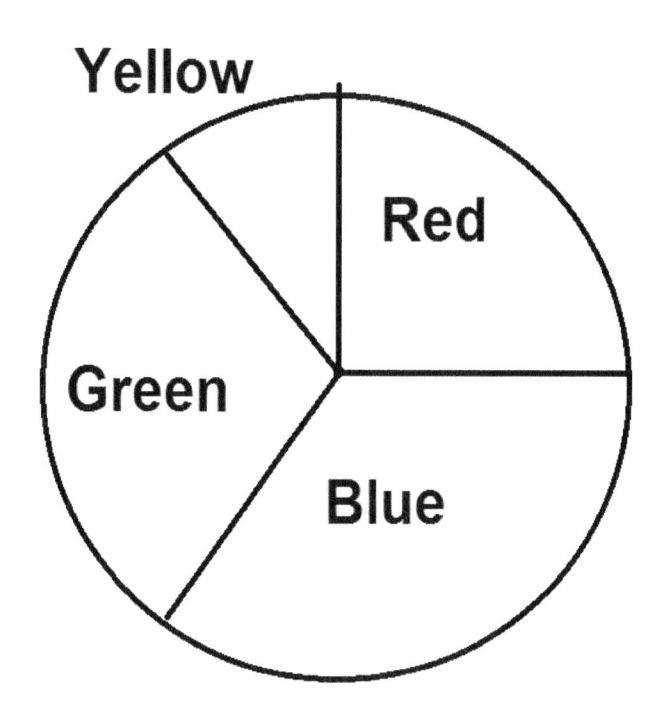

How many times more people said their intention was to vote for the red party than the yellow party?
A. 2   B. 3   C. 4   D. 5   E. 6

Question 21:
A takeaway pizza restaurant is having a sale. If you spend £30 or more at full price, you can get 40% off.

Prices are as follows:
- Basic cheese and tomato pizza: £8 small, £10 large
- All other toppings £1 each
- Sides
    o Garlic bread £3
    o Potato wedges £2.50
    o Chips £1.50
    o Dips £1 each

Ellie and Mike want to order a large pizza with mushrooms and ham, garlic bread, 2 portions of chips and a dip.

Which of these additional items can they order to minimise the price of their order?
A. Small pizza with pineapple and onion
B. Large pizza with mushroom
C. Barbecue dip
D. 4 portions of potato wedges
E. Garlic bread

Question 22:
Sohail is planning to travel between Newcastle and Middlesbrough by car next month. If Sohail does not pass his driving test before his journey, he will not be able to travel on the A1(M), because learner drivers are not allowed to travel on motorways. If he does not travel on the A1(M), he will have to travel on the A19, as this is the only other route connecting Newcastle to Middlesbrough. Thus, if Sohail does not pass his driving test, he will have to travel on the A19.

Which of the following most closely parallels the reasoning used in this passage?
A. If Anna flies from Durham Tees Valley Airport to Sweden, she will have to travel via Amsterdam, as there are no direct flights from Durham Tees Valley to Sweden. However, Anna instead chooses to fly from Manchester Airport, so she will not have to go via Amsterdam.
B. Merdoc the witch is flying on her broomstick, being pursued by a dragon who is currently flying faster than her. If she does not speed up, she will not be able to escape the dragon. If she does not escape the dragon, she will not be able to warn the high council of the impending invasion by the Mithrolites. Thus, if she does not speed up, she will not be able to warn the high council.
C. An oil company is considering purchasing a new processing plant in order to streamline their oil production process and reduce costs by 50%. If they achieve this cost reduction, they will be able to lower their petrol prices by 20%. However, the oil company does not purchase the plant, so this price reduction will not occur.
D. The Food Standards Agency (FSA) is considering implementing new labelling restrictions for genetically modified products. If they do not implement these restrictions, then companies will not have to label foods containing genetically modified produce. If labelling of genetically modified produce is not made compulsory, then environmental groups will protest, because they feel the public should be made aware what they are purchasing. Thus, if the FSA does not implement the new restrictions, environmental groups will protest.
E. In a school sports day, Rebecca is running in the final event, the 100m sprint. She knows that if she runs the sprint in less than 14 seconds, then the red group will not win gold in this event. The points have worked out such that if the red group do not win gold, then Rebecca's group will win the competition. Rebecca runs the 100m in 13.8 seconds, so her group will win the competition.

Question 23:
"The M1 Abrams tank is widely regarded as the most fearsome tank in the world. Highly advanced depleted uranium composite armour makes it difficult to damage from range, whilst a good top speed in excess of 50kmph

and a large fuel capacity make it difficult to catch and contain the tank in an operational context. Whilst the tank does have weak spots that can be exploited at close range, a formidable 122m smoothbore gun as the main armament makes this an incredibly dangerous tactic for opposing tanks. Country X is developing a new main battle tank to boost the prowess of their armoured formations, and have released a statement describing how they will implement next-generation armour into this new tank, to boost its defensive capacity. The government of country X believe this will allow their new tank to compete with the best tanks in the world. However, this view is mistaken. The M1 Abrams clearly demonstrates that a *combination* of different factors, including protection, manoeuvrability and firepower, are responsible for its status as the world's most formidable tank. Simply increasing the defensive capabilities of a tank is not sufficient. Thus, Country X's government is clearly incorrect in this matter."

Which of the following best illustrates the main conclusion of this passage?
A. Increasing the defensive capacity of a tank is not sufficient to make it equal to the best tanks in the world.
B. Multiple factors are required to make a tank equal to the best tanks in the world.
C. The new tank will not be as good as the M1 Abrams, as its defensive capacity will not be as good.
D. The view of Country X's government, that increasing the defensive capacity of a tank will make it equal to the best in the world, is clearly incorrect.
E. No tank is able to compete with the M1 Abrams, which will always be the world's most formidable tank.

Question 24:
The table below shows the balances of my bank accounts in pounds. Interest is paid at the end of the calendar year. My salary, which is the same every month, is paid into my current account on the 2nd of each month. All the money I have is in one or the other of my bank accounts.

	Current Account	Savings	ISA
1st March	1300	5203	2941
1st April	3249	2948	2941
1st May	4398	9384	0
1st June	3948	8292	0

In which month did I spend the most money?
A. February
B. March
C. April
D. May
E. 2 or more months are the same

Question 25:

On Monday, my son developed a disease. No-one else in the house has the disease. The doctor gave me some medicine and told me that everyone in the house who does not have the disease should also take half the dose. We need to take the medicine for 10 days, and the dosage is based on weight.

Weight	Dosage
Under 30kg	0.1ml per kg, 3 times a day
30kg – 60kg	0.2ml per kg, 4 times a day
60kg +	0.1ml per kg, 6 times a day

My son is 40kg. I also have a daughter who is 20kg. I am 75kg and my husband is 80kg. How many 200ml bottles of medicine will we need for the whole 10 days?

A. 4   B. 5   C. 6   D. 7   E. 8

Question 26:
At Tina's nursery school, they have red, yellow and blue plastic cutlery. They have just enough forks and just enough knives for the 21 children there. There are the same number of forks as knives of each colour. Twice as many pieces of cutlery are yellow as blue. Half as many pieces of cutlery are red as blue. Tina takes a fork and a knife at random. What is the probability that she will get her favourite combination, a red fork and a yellow knife?

A. 4/49   B. 1/9   C. 36/49   D. 3/9   E. 3/49

Question 27:
"The UK's taxation and public spending is horrendously flawed, with various immoral features. One example of such a flaw is the subsidisation of public transport with money raised via taxation. According to recent research, public transport is only used by 65% of the population, and since there is no economic benefit stemming from a good public transport system, the other 45% of the population gets no benefit from public transport, but are still required to pay towards it via taxation. The system is in urgent need of reform, such that taxation is only used to support services and systems which are of benefit to everyone."

Which of the following is the best application of the principle used in this passage?

A. Only 48% of the population have ever visited an art gallery, so public funds should not be used to subsidise art galleries, as not all the population use it.
B. Primary and Secondary education provides an economic benefit to the whole country, so public funds should be used to support schools.
C. Although many people never use a hospital, we should still use public funds to provide them, because many people cannot afford private healthcare, and thus we need a publically available health service for those people.
D. There is no evidence that the fire service provides any benefit to the majority of the public, who will never experience a house fire in their lifetime. Thus, the fire service should not be publically funded via taxation.
E. The Police service is a vital service for the country so should be publically funded regardless of how few people benefit from its presence.

Question 28:

"SpicNSpan Inc is a cleaning company offering a range of cleaning services across the UK. The board has recently acquired a new chairman, who has called a meeting of the board to assess how the company can move forwards, expanding its services and increasing its market share. One of the things the new chairman is looking at is the types of services the company provides. He argues that their "All Inclusive" service, where customers pay a fixed amount to have their entire house cleaned as a one-off event, are more popular than their "Hourly" services, where customers pay for a cleaner to carry out a certain number of hours each week. The new chairman argues that they should therefore focus on the "All Inclusive services", rather than the "Hourly" services, in order to increase profits."

Which of the following best illustrates a flaw in the chairman's argument?
A. He ignores other services which may bring in even more profit than 'All Inclusive' services.
B. The fact that 'All inclusive' services are more popular than 'Hourly' services does not mean that they are more profitable. 'Hourly' services may be more profitable.
C. He has assumed that 'Hourly' services are more popular than 'All Inclusive' services.
D. He has assumed that 'All Inclusive' services are more popular than 'Hourly' services.
E. The rest of the board may have other strategies to increase profits, which are better than the new Chairman's.

Question 29:
"The effects of fossil fuels such as Oil, Coal and Natural Gas on the environment are plain and clear for everybody to see. The long-term use of such non-renewable fuels to produce power has led to devastating climate change, and will continue to cause damage as long as it continues. With this in mind, the European Commission has devised a set of targets to promote energy production by alternative types of fuels. However, there is a glaring problem with these targets. Shockingly, the Commission has targeted a "150% increase in the amount of energy produced by Nuclear Power by 2025". This is an outrageous misjudgement, because Nuclear Power is a non-renewable fuel, just like Oil, Coal and Natural Gas. If we wish to protect the environment and halt climate change, we need to switch to *renewable* fuels, which are proven not to cause damage to the environment, NOT non-renewables such as Nuclear Power."

Which of the following best illustrates a flaw in this passage?
A. It has assumed that all non-renewable power sources cause environmental damage.
B. It has assumed that renewable energy sources do not cause environmental damage.
C. It has assumed that the targets will be met, when in fact there is no guarantee that this will happen.
D. It has neglected to consider other problems with the targets set by the Commission.
E. It has assumed that the climate change caused by burning of Oil, Coal and Natural Gas cannot be offset or prevented by other strategies.

Question 30:

"Despite the overwhelming evidence which certifies that vaccines are a miracle of modern medicine, and are responsible for saving a great number of lives, there remains a stubborn section of society that refuses to be vaccinated against important diseases, insisting that they are unsafe and ineffective. This group maintain this view in spite of extremely strong evidence that vaccines are safe, and in spite of the advice given by doctors. This group is particularly strong in the USA, where they pose a very real concern. Over the last 5 years, the percentage of the population that is unvaccinated has been rising by 1 each year, such that now a staggering 6% of Americans have not received any vaccinations. Experts have advised that due to the way diseases are spread, if less than 90% of the population at any given time is unvaccinated, then it is almost certain that we will see an outbreak of Measles, a highly contagious and damaging disease. Thus, we expect that there will likely be an outbreak of measles in the next 5 years in the USA, and we should take steps to prepare for this."

Which of the following, if true, would most strengthen this argument?
A. New and powerful evidence of the safety of vaccinations is due to be released to the public next year.
B. Measles is a highly damaging disease, which frequently causes death or severe permanent injury in those affected.
C. Throughout the last half-century, the number of people who are not vaccinated has risen and fallen continuously. Usually, the increases in non-vaccinated individuals occur over a 6-year period, after which time vaccination becomes more popular, and this number falls.
D. The number of doctors advising against vaccination has been rising for the last 10 years, and shows no signs of decreasing.
E. The rise in unvaccinated individuals has been increasing steadily for 5 years. The only time such a rate of increase has occurred in history was during the 1950s/1960s. In this case, a similar rate of increase in non-vaccinated individuals was maintained for a staggering 13 years.

Question 31:
"It is well established that modern humans evolved in Africa, around 2 million years ago, and that the first humans were mainly hunter-gatherers, living off hunted meat and plant foods collected from their environment. However, this poses an interesting question. Humans are relatively weak, small, feeble creatures, and around 2 million years ago most wildlife in Africa consisted of large, powerful creatures. Thus, it is unclear how humans were able to hunt successfully, and obtain meat for food. One theory is that humans are well-built for long-distance running, largely thanks to our ability to control our temperature via sweating. This theory reasons that humans were able to pursue animals such as antelope, which run when challenged, and were able to keep on running until the antelope collapsed through heat exhaustion. Meanwhile, the humans were kept cool via sweating, and were able to then go in and butcher the defenceless antelope.
Recent evidence has emerged supporting this theory, showing that human feet are well-developed for long-distance running, with fleshy areas in the correct orientation to absorb the impact without causing joint damage, and a heart well evolved to keep pumping at a moderately fast pace for long periods. With the emergence of this powerful new evidence, we should accept this theory, known as "the persistence running theory" as true."

Which of the following identifies a flaw in this argument?
A. The emergence of evidence in support of the persistence running theory does not mean that this theory is true.
B. There is little evidence that the human body is well setup for long-distance running.
C. It has neglected to consider other theories for how humans obtained meat during their early evolution.
D. There are numerous issues with the theory of persistence running, but many of these have been resolved thanks to the new evidence that has emerged.
E. It has not considered evidence that humans evolved in Europe, where there are smaller animals which humans may have easily been able to tackle.

Question 32:

		PREDICTED					
		A	B	C	D	E	U
ACTUAL	A	7	4	2	1	0	0
	B	3	8	2	2	1	0
	C	2	4	5	7	3	1
	D	2	2	2	6	5	0
	E	1	2	2	1	7	2
	U	1	1	0	3	5	6

The table above shows the actual and predicted AS grades of 100 students at Greentown Sixth Form. Assuming that each student can only be predicted one grade, what percentage of students had their grades correctly predicted?

A. 14%   B. 16%   C. 39%   D. 61%   E. 78%

Question 33:
In one year, Mike lowers his workers' wages by x%. The next year, he lowers their wages by x%. The year after this, he raises the wages by x%. In the final year, he raises their wages by x%. In all these stages, x is a constant positive number. Compared to the workers' original wages before any raising or lowering, what are their new wages?
A. The same as the original wages
B. Lower than the original wages
C. Higher than the original wages
D. Can't tell from the provided information even if we know what x was
E. Can't tell from the provided information but would be able to tell if we knew what x was.

Question 34:
"The medical scientific establishment has a long established system for naming body parts and medical phenomena. This system is based upon ease of understanding, such that a body part, or a process of the body, is named based on its clinical relevance. This means that features are named in a way which will help doctors understand and explain to patients what the body part is, or what is wrong with it in the case of a disease. However, this poses significant problems for scientific medical research. Often, the most important features of a body part from a scientific point of view are not the most clinically important features, leading to confusion within the scientific literature, as medical researchers misunderstand the purpose of a discussion, due to confusing nomenclature. Whilst it is important for doctors to be able to explain things clearly to patients, it is relatively easy for this to happen in spite of confusing nomenclature, whereas confusing names causes serious problems in the scientific world. Thus, the naming system for medical features should be altered, to reflect the scientifically important features of body parts, rather than the clinically important ones."

Which of the following best illustrates the main conclusion of this passage?
A. The naming system based on clinically important features causes problems in scientific literature.
B. Changing the naming system would allow faster progress to be made in scientific medical research.
C. The naming system should be changed to reflect the features of body parts which are most important scientifically.
D. The current naming system is sufficient and should not be changed to help lazy scientists who cannot be bothered to do fact-checking.
E. It is more important to have good doctor-patient relations than good progress in scientific research.

Question 35:

Applicants for language teacher training have to specify which languages they studied as part of their degree. 180 people applied for teacher training. Of these, 128 did French as part of their degree. Half as many as did French did Spanish. Three quarters as many as did Spanish did not do either French or Spanish.

How many must have done both French and Spanish?

A. 12   B. 24   C. 36   D. 48   E. 60

Question 36:

"The government has recently had some impressive results from a campaign to reduce drug usage via education about the effects and harm that drugs can cause. However, this strategy fails to tackle one of the major causes of drug usage, namely social deprivation. Many people from deprived backgrounds take drugs due to peer pressure, as they feel that everyone around them is doing it, and that if they do not partake in the drug taking as well, they will be shunned by those around them. Here, no amount of education of the negative side effects of drugs can persuade them not to take them, because the simple fact is that in this situation there are also negative side effects to *not* take drugs, namely social exclusion. Thus the only way to persuade these people to not take drugs is to put more funding into youth centres and recreational activities, thus providing an alternative outlet to these young people, by which they can escape the social exclusion caused by not taking drugs, and engage with a different set of people, who also do not wish to take drugs. Thus, education alone cannot provide any further reductions in drug usage."

Which of the following would most weaken this argument?

A. France has put significantly more money into tackling drugs than the UK over the last few years, but also focuses on education about the negative side of drugs. France has achieved roughly the same reduction in drug usage as the UK thanks to the campaigns.
B. Germany has recently increased spending by over 50% on youth centres and sports groups in areas of high social deprivation, but has had no more success than the UK in reducing drug usage.
C. Countries which do not have significant social deprivation tend to have little issue with drug usage.
D. Ireland had a significant drug problem 15 years ago, which was vastly more challenging than the UK's current drug problem. The Irish government put significant funding into providing sports facilities in areas of high social deprivation, and achieved a remarkable reduction in drug usage.
E. 7 years ago, the Spanish government initiated a wide-reaching campaign, with significant funding to provide education about drugs in primary schools, teaching young children how drugs could damage people's lives. Since the launching of this programme, Spain has seen a reduction in drug usage which far exceeds that seen in the UK.

Question 37:

A touring musical goes to 9 different locations every two months and spends the same number of days being performed in each place. It takes 2 full days to travel between places, i.e. if the last performance in Newcastle is on Tuesday, the musical's first performance in its new location, York, can be at the earliest on Friday. The show is performed every evening, and there are additional matinee (afternoon) performances every Monday, Wednesday and Saturday. Given 1000 people attend each performance, what is the maximum number of people that can see the show during June and July combined in any year?

A. 61,000   B. 64,000   C. 70,000   D. 72,000   E. 80,000

Question 38:

A swimming tournament involves each of 72 competitors doing 3 timed heats. The 216 swims are ranked (i.e. the fastest swim is "1", the second fastest is "2"...) and the ranks of each swimmer's 3 swims are added together to give that swimmer an overall score, i.e.. if Jess swims the 10th, 20th and 50th fastest swims, her score is 80. Swimmers with total scores of 80 or less proceed to the final. If every swimmer swims exactly the same time in each of their swims, and no swimmer swims the same time as any other swimmer, how many swimmers will proceed to the final?

A. 9 B. 10 C. 11 D. 12 E. 13

Question 39:
"The damage that has been wrought on the environment by the use of fossil fuels is well established, and it is widely accepted as essential that we pursue other means of energy production. However, in spite of this, one of the major alternatives to fossil fuels, Nuclear Power, has met with fierce criticism from the public in Europe, most of whom are fearful of it. Some point to nuclear disasters such as Chernobyl and Fukushima as evidence that Nuclear power is not safe, and should not be widely implemented. However, these widely-held fears are not logical, and Nuclear power is perfectly safe. The two disasters can easily be explained: at Chernobyl a misguided procedure was carried out, during which some essential safety protocols were not followed; and the Fukushima disaster was caused by an earthquake and a tidal wave of a magnitude never seen in Europe."

Which of the following cannot be reliably concluded from this passage?
A. The widespread fear of Nuclear Power in Europe is not logical.
B. The nuclear disasters of Chernobyl and Fukushima have caused the widespread fear of Nuclear Power in Europe.
C. The Fukushima nuclear disaster was caused by a tidal wave and an earthquake.
D. Fossil fuels are well established to cause environmental damage.
E. Nuclear Power is perfectly safe.

Question 40:
Bottles of squash contain 1 litre to the nearest decilitre. Megan wants to make 12 litres of a dilute squash mixture which contains exactly 3 times as much water as squash. How many bottles of squash does she need to buy to guarantee she has enough squash?

A. 2 B. 3 C. 4 D. 5 E. 6

Question 41:
Alice and Andrew both go for a run on Sunday morning. In 1 hour, Alice runs 4 times around the park and Andrew runs 3 times around the park. Amanda can only run at 60% of the speed that Alice can. Next week, Andrew and Amanda both run a 10km race. Amanda runs the race at a speed of 200 metres per minute. Assuming they both run at the same constant speed they do in training, how long in minutes and seconds does Andrew take to run the race?

A. 40:00 B. 42:30 C. 44:50 D. 46:00 E. 48:20

Question 42:
"The UK public should accept an increase in their national insurance contributions. Nobody wants to pay more taxes, but it is a fact that the population is ageing. Elderly people are more prone to health problems, meaning that the NHS requires more funds to deal with this extra workload. Given the choice, the vast majority of the people in Britain would choose having a properly funded NHS over having lower taxes."

Which of the following best illustrates the main conclusion of this passage?
A. Nobody wants to pay more taxes.
B. The vast majority of the public would choose having a properly funded NHS over having lower taxes.
C. Elderly people have more health problems.
D. The UK has an ageing population.
E. The UK public should accept an increase in their national insurance contributions.

Question 43:

	Pool A	Pool B	Pool C	Pool D
1st	France	Argentina	England	South Africa
2nd	Holland	Mexico	Nigeria	Brazil
3rd	United States	Denmark	Germany	Japan
4th	India	Korea	Ghana	Algeria
5th	Australia	Switzerland	Portugal	Serbia
6th	Greece	New Zealand	Honduras	Uruguay
7th	Chile	Slovakia	Cameroon	Paraguay

The table above shows the final standings in the pool stages of a football competition. The top 2 teams from each pool progress into the quarter-finals. The fixtures for the quarter-finals are determined as follows:

QF1: Winners Pool A vs. Runner up Pool B **QF3:** Winners Pool C vs. Runner up Pool D
QF2: Winners Pool B vs. Runner up Pool C **QF4:** Winners Pool D vs. Runner up Pool A

The winners of QF1 then play the winners of QF3 in one semi-final, and the winners of QF2 and winners of QF4 play each other in the other semi-final. The winners of the semi-finals progress to the final.

Which of these teams could England play in the final?
A. Nigeria   B. France   C. Mexico   D. Denmark   E. Brazil

Question 44:
Kelly is working on a school project. She begins by making a large card surface on which to display her project by attaching A4 (300mm x 210mm) sheets of card together with tape. She applies tape to the entire back and front of every join she makes. She wants her large card to be at least 1 metre by 1 metre and doesn't cut or overlap any card.

How much tape does she need in total to make her large card surface?
A. 3.51m   B. 5.85m   C. 7.02m   D. 10.7m   E. 107m

Question 45:
Katie's netball team have played 24 matches this season. They play each team once at home and once away. In total they have won 18 matches. They have won twice as many matches at home as away. They have not drawn any matches. How many more matches have they lost away than lost at home?
A. 2   B. 3   C. 4   D. 5   E. 6

Question 46:
"Science fiction movies have a lot to answer for. In the UK, many Sci-fi movies have contributed to a remarkably poor understanding of science, and in some cases, a misguided and illogical fear of scientific progress. One prime example is the genre of zombie horror, which frequently features a virus taking over people, modifying their behaviour in a matter of seconds, and turning them into deranged, mindless zombies. Such a phenomenon is impossible; the closest any virus comes is the rabies virus, which takes several days at the minimum to stimulate aggressive, raging behaviour. Such illogical fear often leads to less public support for science funding, which severely hinders research into cures for many devastating diseases. Thus, it is clear that without science fiction movies, many deaths could be prevented."

Which of the following best illustrates a flaw in this passage?
A. The example of zombie horror movies may not be applicable to other sci-fi movies.
B. There is no evidence that any viruses have the ability to infect people and modify their behaviour within a few seconds.
C. It has assumed that if research on diseases were not hindered, many deaths could be prevented.
D. It has assumed that there are no other contributing factors to public fear of scientific progress.
E. It has assumed that the public fear of science is illogical.

Question 47:

Ashley, Ben, Callum, Dave and Ed agree to meet in the city centre at 1pm. Ashley walks from his house 8km away at a speed of 8km per hour. Ben gets the bus, which takes 40 minutes and departs from his house at 25 and 55 minutes past the hour, every hour. Callum cycles 12.5km at a speed of 12km per hour. Dave gets the train, which goes every 10 minutes and takes 20 minutes, but he has to walk to the station which takes 25 minutes. Ed drives to the park and ride which takes 10 minutes, then gets the park and ride bus which comes every 10 minutes and takes 15 minutes.

Who has to leave their house the earliest to get to the city centre on time?
A. Ashley   B. Ben   C. Callum   D. Dave   E. Ed

Question 48:
6 friends from university all send each other Christmas cards. Posting each card costs £0.50, apart from cards to and from Sophie, who lives abroad. Posting cards abroad costs £1.50, and sending cards from abroad costs £1.20.

How much in total is spent by the 6 friends on sending cards?
A. Between £20 and £21.99
B. Between £22 and £23.99
C. Between £24 and £25.99
D. Between £26 and £27.99
E. Between £28 and £29.99

Question 49:

	Goals scored For	Goals scored Against
City	10	4
United	8	5
Rovers	1	10

The table above shows the goal scoring record of teams in a football tournament. Each team plays the other teams twice, once at home and once away. Here are the results of the first 4 matches:
United 2 – 2 City   City 2 – 1 Rovers
Rovers 0 – 3 City   Rovers 0 – 3 United

What were the results of the final two fixtures?
A. United 2 – 0 Rovers, City 0 – 0 United
B. United 1 – 0 Rovers, City 1 – 1 United
C. United 0 – 0 Rovers, City 2 – 1 United
D. United 1 – 0 Rovers, City 2 – 2 United
E. United 2 – 0 Rovers, City 3 – 1 United

Question 50:
"The UK has recently had an election, and the opposition won, meaning a new cabinet has been formed. The new education minister has argued that the UK should introduce more coursework into GCSE and A Level courses, because it is a much fairer way to assess students than via examination. In fact, coursework does not measure a student's ability in a subject, but how much help they receive outside their lessons. Students with help from relatives or friends who are knowledgeable in certain subjects are shown to perform much more strongly than others with a similar ability but less help. Thus, coursework is clearly not a fairer method of assessment than examinations, so the new education minister is clearly incorrect."

Which of the following best illustrates the main conclusion of this passage?
A. The new education minister is not a logical person.
B. Coursework is not the fairest method of assessment for GCSE and A Level courses.
C. Coursework measures students on the level of outside help they receive, not their ability in a subject.
D. The education minister's argument is incorrect.
E. There are no methods of assessment fairer than examination.

END OF SECTION

YOU MUST ANSWER ONLY ONE OF THE FOLLOWING QUESTIONS

Question 1:
To what extent are 'logical' and 'rational' synonymous?

Question 2:
In what instances is aggression justified?

Question 3:
What are the limits of scientific theories of human behaviour?

Question 4:
Assuming time travel was possible, could we learn more from the past or the future?

END OF TEST

ANSWERS

Answer Key

Paper A	Paper B	Paper C
1 B	1 C	1 D
2 D	2 B	2 A
3 E	3 B	3 C
4 C	4 B	4 D
5 C	5 D	5 A
6 C	6 B	6 B
7 E	7 D	7 E
8 C	8 C	8 E
9 D	9 B	9 D
10 D	10 C	10 B
11 C	11 E	11 A
12 C	12 A	12 D
13 D	13 C	13 C
14 B	14 B	14 A
15 B	15 C	15 E
16 E	16 E	16 D
17 C	17 D	17 B
18 C	18 E	18 B
19 C	19 A	19 B
20 B	20 D	20 B
21 C	21 E	21 B
22 E	22 E	22 D
23 C	23 D	23 D
24 C	24 B + E	24 D
25 B	25 D	25 B
26 D	26 C	26 A
27 D	27 B	27 D
28 E	28 C	28 B
29 B	29 C	29 A
30 C	30 C	30 E
31 C	31 A	31 A
32 D	32 A	32 C
33 C	33 D	33 B
34 B	34 C	34 C
35 D	35 B	35 E
36 B	36 E	36 E
37 C	37 C	37 D
38 D	38 B	38 A
39 D	39 C	39 B
40 E	40 E	40 C
41 B	41 D	41 A
42 D	42 C	42 E
43 D	43 C	43 A
44 D	44 B	44 D
45 B	45 A	45 E
46 C	46 B	46 C
47 C	47 C	47 B
48 E	48 C	48 B
49 A	49 C	49 E
50 B	50 B	50 D

MOCK PAPER A SECTION 1 — ANSWERS

Raw to Scaled Scores

1	10.5	11	38.5	21	49.5	31	59.0	41	70.5
2	18.0	12	40.0	22	50.5	32	60.0	42	72.0
3	22.5	13	41.0	23	51.5	33	61.0	43	74.0
4	25.5	14	42.0	24	52.5	34	62.0	44	75.5
5	28.5	15	43.5	25	53.5	35	63.0	45	78.0
6	30.5	16	44.5	26	54.5	36	64.0	46	80.5
7	32.5	17	45.5	27	55.0	37	65.0	47	83.5
8	34.0	18	46.5	28	56.0	38	66.5	48	88.0
9	35.5	19	47.5	29	57.0	39	67.5	49	95.5
10	37.0	20	48.5	30	58.0	40	69.0	50	103

Mock Paper Answers

Mock Paper A: Section 1

Question 1: B
James runs 26.2 seconds, which is outside the qualifying time, therefore he does not qualify

Question 2: D
5.6/7 gives the unit price of 80p – this equals a packet of crisps. Multiplying this by 2 gives the sandwich and by 4 gives the watermelon price of £3.20

Question 3: E
Jane leaves at 2:35pm and arrives at 3:25pm, taking 50 minutes. Sam's journey takes twice as long, so leaving at 3:00pm it takes 100 minutes, giving an arrival time of 4:40pm

Question 4: C
After the donation, Michael has eight sweets. Therefore, Hannah had 16 sweets after the transaction and hence 13 sweets before

Question 5: C
Find original pay: £250/0.86 = 290.697674 basic original pay. Add the rise: (290.697674 x 1.05) + 6 = £311.232558 new basic pay. Subtract the income tax at 12% = 311.232558 x 0.88 = £273.884651 new pay rate. To the nearest whole pound this is £274

Question 6: C
Given the first cube is a white cube, you are drawing from one of three boxes, boxes A, C or D. Boxes C and D will have just had their only white cube removed, whereas box A will have one white cube remaining. Therefore, the probability of drawing a second white cube is $1/3$, thus the probability of non-white (i.e. black) is $2/3$.

Question 7: E
This is a simultaneous equations question. $500 + 10(x - 80) = 600 + 5x$; true when $x \geq 80$.
$500 + 10x - 800 = 600 + 5x$
» $5x = 900$
» $x = 180$

Question 8: C

MOCK PAPER A SECTION 1 — ANSWERS

If eating more slowly caused a reduction in the time available to work, the candidate might be less productive.

Question 9: D
This is a LCM question. We need to find the lowest common multiple of the song lengths. The LCM of 100, 180 and 240 is 3,600 seconds – equal to 60 minutes. For ease of arithmetic, you may choose to work reduce all numbers by a factor of 10.

Question 10: D
The journey is 3 hours and 45 mins, minus a 14 minute break gives 3hrs 31 mins travel time, or 211 minutes. Therefore, the average speed is 51mph, or 82kph by using the stated conversion factor.

Question 11: C
The mean guess is £13.80, which is £5.80 too high

Question 12: C
The overall error for respondent 3 is £13, which is the least

Question 13: D
Scale back and forth from known quantities. Country B has 32.1m so Country D has 38.5m people.

Question 14: B
Country B has 32.1m people. Therefore $45 x 32.1m = $1.44bn

Question 15: B
The average speed is 24mph, independent of distance travelled as it cancels. Imagine this covers a set distance of say 30 miles. It will take 1 hour on the way and 1.5 hours on the way back. 60/2.5 = 24. This is true of all distances, the ratio is the same.

Question 16: E
None of the above can be reliably deduced from the passage alone

Question 17: C
Imagine the toothpaste costs 100p originally, and follow the price through. It rises by 80% to 180p, then is reduced by 50% to 90p. Three tubes are purchased for the price of 2 (i.e. 180p), therefore the cost per unit is 180/3 = 60p. 60p = 60% x 100, the original price'

Question 18: C
Argument C has the same form, asserting that since something is not happening, the result of the action will never be true.

Question 19: C
Statement C is the only one making reference to the potential outcome of solving crimes faster, thereby providing a plausible mechanism for a reduction in cybercrime rates

Question 20: B
The passage suggests that the attacks were carried out by extra terrestrial beings. Though the supposed UFO sightings have rational explanations, the writer feels this is insufficient to dismiss his idea.

Question 21: C
The initial argument suggests that two things must be present for an action to happen. If only one is absent, the action cannot happen. Argument C has the same form, the others do not.

MOCK PAPER A SECTION 1 — ANSWERS

Question 22: E
Growing vegetables needs several positive traits. The passage does not tell us which is the most important or most commonly lacked skill, only that more than one skill is required for success.

Question 23: C
Joseph does not have blue cubic blocks, since all his blue block are cylindrical.

Question 24: C
130°. Each hour is 1/12 of a complete turn, equalling 30°. The smaller angle between 4 and 8 on the clock face is 4 gaps, therefore 120°. In addition, there is 1/3 of the distance between 3 and 4 still to turn, so an additional 10° must be added on to account for that.

Question 25: B
The chance of red is 2/6 = 1/3. To get no reds at all, it must be non-red for each of three independent rolls. The probability of this is $(2/3)^3 = 8/27$. Therefore the probability of at least one red is $1 - 8/27 = \underline{19/27}$

Question 26: D
These three furniture items are compatible with having 6 legs. All the other statements are false.

Question 27: D
Work this out by time. The friends are closing on each other at a total of 6mph overall, therefore the 42 miles take 7 hours. In seven hours, the pigeon, flying at 18moh covers 18 x 7 = 126 miles.

Question 28: E
The passage does not make any supported claims about fruit juice. It gives rationale for both benefits and risks of fruit juice consumption without reaching a conclusion.

Question 29: B
Calculate the overall cost of three stationery sets, then subtract any items not bought. For each item shared between two people, there is one of that item not required. The overall cost is £6.00 per person, £18.00 overall. Subtract one geometry set (£3), one paper pad (£1) and one pencil (50p) to give £13.50 overall cost.

Question 30: C
Argument C is the most convincing. It gives a strong rationale as to why the notion that people should only pay for services which they personally use is likely to have serious adverse consequences on the nation as a whole. Therefore this flawed logic is not suitable to apply to the arts funding dilemma.

Question 31: C
Moving matches 1 and 4 to form a cross inside one of the other cubes will solve the problem. Two squares are broken (the top left hand corner and the overall large square) but four new small ones are created, bringing the total up to seven.

Question 32: D
The white square is opposite brown, since both are adjacent to blue on opposite sides. White and brown cannot be adjacent to each other since the position of the opposite black and red sides makes that impossible.

Question 33: C
We take the overall price to the UK and subtract money which does not go to the farmers. 36,000,000kg at 300p/kg gives £108m. Subtract commission 108 x 0.8, then take 10% of the remaining proceeds as the farmers' share, giving £8.64m

Question 34: B
The first paragraph tells us annual road deaths have fallen, so B is true. The others are false.

MOCK PAPER A SECTION 1 — ANSWERS

Question 35: D
Regression to the mean is a phenomenon observed when a value is variable within a probability distribution. Sometimes by chance it will be at the high or low end, but thereafter it is likely to be closer to what is expected. This can explain the fall in drink driving deaths after the new campaign.

Question 36: B
The relevant set is 'people with a sore throat and a chest infection'. If *some* (i.e. not *all*) members of that set have the 'flu', then it follows, necessarily, that the other members of the set do not. This is because you can only either *have* the 'flu', or *not have* the 'flu'. So, B is supported.

Question 37: C
Catherine must choose four socks. If choosing three or fewer, it is possible that they could each be of different colour. When choosing four, it is certain that at least two socks will make a matching pair, but possible that there will be two pairs.

Question 38: D
This is another simultaneous equations question. Solve to find x, the normal rate of pay.
$100x + 20y = 2000$ » $60y = 6000 - 300x$ (substitute this)

$80x + 60y = 2700$
» $80x + (6000 - 300x) = 2700$
» $220x = 3300$
$x = 15$
Substitute $x=15$ into one of the original equations, and then solve to get $y = 25$

Question 39: D
The easiest way to do this is via simultaneous equations. Let A be the distance travelled by the Plymouth train and B the distance travelled by the Manchester train. Thus:
$A = 90x + 45$ and $B = 70x$
The collision will occur when the total distance travelled by both trains is = 405
i.e. $A + B = 405$
Therefore, $90x + 45 + 70x = 405$
$X = 2.25$ hours. Thus, collision happens at 12:45.
Substitute $x=2.25$ into the first equation to give the distance from Plymouth:
$A = 90 \times 2.25 + 45 = 247.5$, which rounds to 248 miles

Question 40: E
Statement E is not true, the others are true. A pregnant rabbit requires 50 pieces per day and a normal rabbit requires 25. Therefore three pregnant and ten normal rabbits require only 400 pieces per day, not 450.

Question 41: B
If memory of names uses a different part of the brain, then conclusions drawn from this experiment may have no validity.

Question 42: D
Michael pays £60 and £110 = £170 for the painting. He sells it for £90 and £130 = £220. Thus, he makes a profit of £220 - £170 = £50.

Question 43: D
The principle problem is that it does not compare the relative effectiveness of pesticides and natural predators. It might be that pesticides are far more effective at controlling pests, despite the unnecessary excess killing.

Question 44: D
Proportionately, there would be 172 members. Therefore there is an excess of 298 – 172 = 126 members.

MOCK PAPER A SECTION 1 — ANSWERS

Question 45: B
If each pair of opposite faces is painted one colour, this requirement can be satisfied with a minimum of three colours.

Question 46: C
Two thin people cross dropping one (first crossing). One returns (second) and then takes another thin person over(third). When one returns this time (fourth) he leaves two on the correct bank and returns for the fat person alone on the wrong side. The fat person comes back alone leaving one thin person on the wrong side (five) and one of the thin people (six) return to collect the final thin person. These two return together completing the 7th crossing.

Question 47: C
Christmas day is a Wednesday. If there are only four Saturdays and four Wednesdays, the 1st December is a Sunday and thus the 25th is a Wednesday.

Question 48: E
Iver is West of Ruddock, but we don't know where it is in relation to Dell as this is not specified, therefore E cannot be concluded. The other statements are true.

Question 49: A
The passage argues that as most women's needs have changed, the style of clothing should change. This is derived from the principle that the needs of the majority should be prioritised.

Question 50: B
This argument has the same structure. It says since one thing is not possible, something more extreme than it with regard to the same characteristic is also not possible. The other arguments have a different construction.

END OF SECTION

Mock Paper A: Section 2

1. *"Strive not to be a success, but to be of value"*
   To what extent is it possible to be "a success", but to have little value?

Introduction:

- Definitions of success and value should be given, with a brief exploration of what is implied by the two terms and any connotations.
- Success could be considered to be defined through achievements or status, whilst value is defined on a more emotional or social level with the influence that people have on the situation around them.
- The introduction should begin to explore where there are overlaps between the two terms and where they could be seen as standing alone. This could be done by general statements or through specific examples, so long as valid ideas are introduced on both sides.

Possible arguments for success without value:

- In the context of money – a person could have a successful career defined by the amount of money that they make, but may not have been a positive influence on those around them. You could consider the extent to which the definitions may get confused with worth but doesn't mean it (e.g. she's worth £xxx as a term for success)
- In the context of academic achievement – just because you have been successful with grades at school, for example, doesn't mean you were necessarily valued in the setting or by those around you.
- In the context of gambling or competitive sports/games – just because a person has been successful in 'winning' doesn't necessarily mean that they were valued as a player or by others around them.

Possible arguments for success and value being integrated:

- Is success a value in itself? Could this be in the eye of the beholder?
- In terms of gaining power or authority – you could choose to explore whether figures in power such as a manager, politician etc. must be valued by somebody to reach this status?
- In particular careers such as teaching or charity work – in order to be successful must you be doing something that is valued by somebody?
- On an emotional level, could you be considered a 'successful' friend or family member if you are valued within this setting? This may be particularly relevant using the example of a mother figure.
- You may wish to put forward the idea that you can be valuable without success, but does this work the other way round?

Conclusion:

- Include a summary of all points given, and refer back to your original definition.
- The conclusion should reach a clear and logical solution – you could choose to completely decide that it is impossible, but most probably will reach a compromise which encompasses the idea that value is distinct from success, but the terms can be used interchangeably in particular situations.

2. *Is the media a positive or negative influence on scientific understanding?*

Introduction:

- Introduce or illustrate the ways in which the media are involved in our scientific understanding – you could describe what ways we might have access to the information, such as newspapers, TV.
- You may give reference to the particular areas of scientific interest in the media, such as health and global warming.
- Introduce both sides of the argument – the possible positive influences and possible negative ones (ideas below).

Possible positive influences:

- Action- It could be argued that we wouldn't be aware of some of the biggest scientific issues needing tackling if it wasn't for the involvement of the media, and therefore the effect must be positive if we are becoming aware of pressing problems (e.g. obesity)
- Access - The media is a way through which the world of scientific research integrates with common people, who would not otherwise recognise the importance of science.
- Description – the media may be able to explain ideas in a more easily understandable manner, such as by identifying the key trends and emphasising the important facts, which makes science more understandable to those with less knowledge.
- It could be argued that any awareness of science is positive for those who would not have any interest otherwise.
- Examples of specific situations – e.g. in the banning of CFC sprays.

Possible negative influences:

- Simple facts may get blown out of proportion or taken out of context to such an extent that they could no longer even be considered factual or valid.
- False or flawed statistics are often given, taken from unreliable sources, which lead to confusion and the wrong message being portrayed.
- Oversimplification of ideas – everything is dumbed down and will not give the full story.
- Can create a biased public opinion – a conclusion may be reached which does not consider all particular explanations for a phenomenon.

Conclusion:

- Include a summary of all points made, and give a balanced overview of both sides of the argument.
- You may wish to reach a unanimous decision on the effect, making it clear that this is a personal opinion. Or you may wish to come to a compromise where the media are a positive in some situations or to some extent, but after a particular point or when taken too far, this effect can become negative.

3. *"Why tell the truth if a lie is better for all concerned?"*
   In what circumstances can dishonesty be justified?

Introduction:

- It may be important to define the term 'lie' as a statement or implication that is directly opposed to the truth. It is also important to note that, in this context, a lie is a deliberate dishonest remark (not an accidental incorrect statement)
- You may wish to distinguish between lying as deliberately telling something that is factually incorrect, and deliberately avoiding or obscuring the truth.
- Introduce the idea that dishonesty is considered immoral and 'wrong' – the basic social rules are not to lie, children are taught from a young age that lying is bad behaviour.
- Begin to introduce you arguments in favour of lying in particular situations – the points you wish to explore later.

You may wish to structure this essay slightly differently to this mark scheme; through exploring a circumstance where it may be considered justifiable to lie, and then look at the alternative perspective arguing that it is not justified. Alternatively, it would be just as effective to make points for and against, as below.

Possible arguments for:

- In the case of 'white lies' – to avoid discouraging or offending people e.g. "Your hair looks nice"
- In self defence or when lying may save a number of lives – e.g. when people are taken hostage for their beliefs/ethnicity/race
- When talking to children about situations they wouldn't otherwise understand, especially when they ask difficult questions.
- To avoid an unnecessary, difficult conversation with someone who doesn't need to know something detailed – e.g. if they ask, "How are you?" not giving them a truthful answer may avoid a difficult conversation.

Possible arguments against:

- Creates a lack of trust and a lack of stability, feelings of guilt on both sides.
- Religious/foundations – 'do not lie' is one of the Ten Commandments.
- You could argue that lying will always cause more problems later down the line, even if making things easier in the short term.
- Advocating lying in some situations makes it easier to lie in other situations and creates a perpetual problem, it's impossible to draw a line other than to just say one should not lie as a rule.

Conclusion:

- Include a summary of all points made, and give a balanced overview of both sides of the argument.
- You need to reach a decision which illustrates if, and when, dishonesty can be justified. It's OK to also reach the conclusion that it never can be, so long as this is logically stated.

4. *"Science is a nothing more than a thought process"*
   What actually is science and how is it of value to us?

This essay is very open-ended: it is possible to take this question in any direction you feel appropriate depending on personal knowledge and interest so this mark scheme is by no means the only way of approaching this question – so long as a convincing and balanced argument is given and both elements of the question are considered, the answer will be credible.

Introduction:

- Include a definition of science – a systematically organized study or body of knowledge of the physical and natural world.
- Begin to introduce some of the key features of science.
- Begin to introduce some of the values that science has for society.

Possible points to consider – you want to contain a balance of defining and exploring the key features of science, and exploring the value of science within our society:

- Science is experimental – involving accurate and detailed study.
- Science could be argued to be physical and measurable- using known concepts in the real world that can be objectively identified.
- Science is adaptable and ever changing – according to research and current thought.
- Science could be considered to be held within the thoughts and brains of society – if there weren't people exploring it or interested in it, would it really exist?
- Science involves logical and critical thought – in that sense it could be considered to be merely a thought process – but you may wish to explore the idea that it couldn't exist without physical things to measure in themselves.
- Valuable for the evolution of society – as we learn more about the world and about ourselves, this helps change to occur which is vital.
- Provides concrete fact and stability.
- Gives a purpose to life, which brings fulfillment and contentedness.
- Brings explanation for some things that can't be explained through simple observations, which allows a firmer foundation and an answer to bigger, crucial questions.
- Helps us to gain more of an understanding of the world we are in – bringing a stronger grasp on reality and greater knowledge and insight.
- Medical reasons – drugs, medicines, health.
- Safety – e.g. monitoring volcanoes, predicting earthquakes etc.

Conclusion:

- Summarize the key ideas explored previously – the ideas associated with science and the values it may hold.
- Ensure to link back to the original statement – is it within our minds and merely a thought process – and does that diminish its' value?

END OF PAPER

Mock Paper B: Section 1

Question 1: C
This question has to be worked through in stages. To begin with, adjust his weekly pay to an annual salary, (560 x 52 = £29,120). Accounting for tax, his annual salary is 29,120/0.9 = 32,355. To account for the pay cut, 32,355 x 0.95 = £30,737. To deduct tax, subtract 20% of 20,737 (his taxable income) to give 4,147 tax. Subtracting the new tax from his new overall salary gives 26,589. Add on the new bonus of 90 per month to give £27,669, or £27,700 when rounded.

Question 2: B
The total saving on the final booking relative to the first is £230, but the cost of two cancellations must be deducted (£90) giving a total saving of £140.

Question 3: B
There are originally no odd numbered balls in Bag A. But as a result of the transfer, there could be an odd ball in Bag A. Therefore the probability of drawing an odd ball is found by multiplying the probability of selecting the new ball ($1/5$) by the probability that that ball is odd ($2/5$ – given by adding the one odd ball in the bag originally to the odd ball introduced) giving an probability of $2/25$ that the selected ball from Bag A is odd.

Question 4: B
Assume the price of bread is 100p. 100 x 1.4 x 0.8 = 112p after the subsidy. The cost of three loaves is therefore 336p (divided four ways this equals 84p per loaf)

Question 5: D
At 2120hrs, the minute hand is pointing to 4 and the hour hand is pointing one third of the way past 9 towards 10. $360°/12 = 30°$ – this is the number of degrees per hour division. Between the two hands then, there are 5 hour divisions plus an extra $1/3$. Therefore the angle is $(30 \times 5)+(30/3) = 160°$

Question 6: B
There is a 3l and 5l bucket – therefore 4 litres can be measured from the difference between the buckets as follows. Fill the 5l bucket, decant 3l into the smaller bucket and then you are left with 2l in the large bucket. Pour this into the tank. Repeat the process again, decanting the remaining 2l into the tank once again to make 4l in total. The first time, 5 litres was required. The second time, the 3 litres from the second bucket could be tipped back into the 5l bucket, and then filled up with 2l of fresh water to then measure the final 2l in. Therefore 5 + 2 = 7 litres of water is sufficient to fill the tank with 4l.

Question 7: D
To answer this question, make a timeline showing the locations of the different genres of books. Place each book on the timeline as appropriate, making sure to indicate where more than one location is a possibility. From that, you will see that literature books are located to the right of engineering. This is true since they are to the right of art (which we know is right of mathematics (and therefore engineering, since the run between the sciences is uninterrupted)). The other statements, whilst potentially true, cannot be deduced for certain.

Question 8: C
The passage tells us that brand new cars lose value quickly, despite the car being virtually unchanged. Therefore in the absence of any contradictory information, it is reasonable to conclude that buying second hand cars is a wise choice.

Question 9: B
First, calculate how many bottles are sold. 2000 – (2000x0.9x0.8) = 560 bottles. Then divide the total profit by the number of units to give the profit per unit, which comes to 11200/560 = £20 per bottle.

Question 10: C
If ocean exploration has led to the discovery of many useful drugs, it could easily be said that it has benefitted many people in trouble. Whilst it might be cheaper than space exploration, the two are entirely different, and therefore people's views on the cost-effectiveness of space exploration cannot be directly compared to their views on ocean exploration. The other responses do not address the effect on people in trouble.

Question 11: E
The definition of timelessness requires something to be tested by time. Something that modern furniture cannot fulfil. Therefore statement E expresses a significant flaw in the reasoning. The other statements do not refer to the 'timelessness' aspect of furniture, therefore they are not directly relevant to the argument.

Question 12: A
The passage talks about the benefits of drinking red wine, not about living near to vineyards. The passage does not state that Italians drink more wine than Germans, therefore the assumption that they do is central to the argument.

Question 13: C
Tom arrives at 1620, and leaves 45 mins after Jane leaves. Therefore he also leaves 45 mins after Hannah leaves, since Jane and Hannah leave together. Since his journey is 10 mins faster than Hannah's, he arrives only 35 minutes after Hannah arrives (which happens to be 1620). Therefore Hannah arrives 35 minutes earlier than this, at 1545. Since she left at 1430, her journey took 75 minutes. Jane's journey took 40% longer (1.4 x 75 = 105 minutes). Therefore leaving at the same time as Hannah, 1430, Jane arrived 105 minutes later at 1615.

Question 14: B
This is a simultaneous equations question. Let **x** be the number of standard tickets sold, and **y** be the number of premium tickets sold.
Therefore: $x + y = 600$; $10x + 16y = 6,600$
$x = 600 - y$ » substitute: $10(600 - y) + 16y = 6600$
$6y = 600$; $y = 100$, therefore 100 premium tickets were sold.

Question 15: C
Between 20th January and 23rd April, there are 94 days. In 94 days, the moon makes 94/28 = 3.36 orbits. This is equal to 3.36 x 360° = 1210°

Question 16: E
In question 14, you are looking for a strong opposition to the proposition that students at drama academies are not taught well academically. The strongest opposition would be evidence that such students perform academically well in some objective measure. Evidence of significantly above average GCSE results provides this.

Question 17: D
You should definitely draw this one out on paper. Trace out the paths and you find that both people have a net displacement of 11km to the North. Therefore since Anil is only net 2km East, and Suresh is 17km East of the starting point, there is a 15km separation between them

Question 18: E
If three times the final amount of concrete is ground off by the builder, three quarters of the original thickness is removed, hence one quarter remains. 14/4 = 3.5cm

Question 19: A
Walking at 4mph, 3 miles takes ¾ hour = 45 mins. Adding the 5 minute stop, Chris will arrive at 1820, since he set off at 1730. At 24mph, 6 miles takes ¼ hour, 15 mins. Therefore setting off at 1810, Sarah will arrive at Laura's at 1825. Therefore Chris arrives 5 minutes earlier than Sarah.

MOCK PAPER B SECTION 1 — ANSWERS

Question 20: D
The passage tells us that illegal downloads are causing harm to the music industry. Whilst it gives an example, this does not mean the stated example is the principal issue. The conclusion that best fits the passage as a whole is to say illegal downloading is more harmful than many people think, given their willingness to undertake it.

Question 21: E
First, calculate the amount of water needed for each type of fire. Typical house fires require 20,000 litres, whereas garden fires usually need only 10,000 litres. Therefore all statements are correct EXCEPT E. Three house and ten garden fires require 160,000 litres to extinguish, not 140,000.

Question 22: E
Calculate the distance travelled during each component of the journey, then add them together. (20 x 30 = 600m, (30 + 20)/2 x 5 = 125m, 30 x 20 = 600m, 30/2 x 10 = 150m). Adding the distances together gives 1475m.

Question 23: D
The passage only talks about people's opinions on the scheme, and not about any action which could potentially be taken. Therefore the best summary is to say that more people oppose the scheme than support it.

Question 24: B + E
The question asks for two responses, therefore you must mark two and get them both correct for one mark. The suggestion is made that reducing fishing will improve fish populations. This assertion carries two major assumptions – that the fishing originally caused the decline, and that the decline is reversible, and can therefore recover if the threat is removed. Select these two responses for a mark.

Question 25: D
To calculate this, you need to work out how many possible combinations there are, and how many of them contain exactly two heads. Since there are 2 possibilities and 5 trials, the number of potential outcomes is $2^5 = 32$. For two heads, any combination of two coins can show heads – and since there are 5 coins tossed, there are 10 possible combinations of exactly two heads. Therefore the probability is $10/32$, which is equivalent to $5/16$.

Question 26: C
For a 5 litre bucket with a 2% margin for error, the maximum possible volume is 1.02 x 5 = 5.10l, and the minimum is 0.98 x 5 = 4.90l. Therefore there is a 200ml difference between the maximum and minimum volume possible. Therefore the range of cleaning powder required is 0.2 x 40 = 8g.

Question 27: B
To calculate the cost of the call, you need to first work out its duration in minutes and multiply by the off-peak rate per minute. Then you add on the connection fee. A call of 1.4 hours = 1 hour 24 minutes = 84 mins. (84 x 22 =1848p), adding the connection fee of 18p gives 1866p, or £18.66

Question 28: C
The passage tells us about the risks of sunbathing, and that many people do not see the danger in it. The final sentence shows us that the conclusion is a wide-ranging one, not a specific observation about UV radiation. Therefore Answer C is correct, it best sums up the passage as a whole.

Question 29: C
The sample argument gives a premise which represents a problem. There are two possible solutions, and since one is not available then the other solution is sought. Argument C has the same structure. The clothes need to be dried, and since the tumble dryer is out of action, the clothes are pinned on the washing line.

MOCK PAPER B SECTION 1 — ANSWERS

Question 30: C
First calculate the total pay, then divide this by the number of hours Jim works for an hourly rate. Total pay = 3 x 8 = £24; Total time = (11 x 8 x 2) + (15 x 7) = 281 minutes = 4.68 hours. 24/4.68= £5.12 [the total time is equal to the number of windows in total multiplied by the time taken to clean each window, plus the time travelling between the houses, which is 15 multiplied by the 7 journeys required]

Question 31: A
The passage argues that bottled water is pointless, as is almost identical to tap water. If bottled water had an additional benefit, such as being good for health, it might be that it makes sense to drink bottled water.

Question 32: A
Be careful – this sentence contains a triple negative. If the sentence read "...nor any cyclists that are marathon runners", it would be clear that no cyclists also run marathons. Changing the sentence to "...nor NO cyclists that AREN'T marathon runners" introduces a double negative, hence the meaning is not changed. Therefore it still means no cyclists run marathons, hence **A** is true.

Question 33: D
Draw this one out on a line. You will see that whilst we know Oakton is East of Langham, we cannot conclude its whereabouts in relation to Frampton – it could be either East or West. Therefore **D** cannot be said with certainty.

Question 34: C
Firstly calculate the surface area of the dome, then divide by the surface area covered by one pot to calculate the number of pots needed. A dome is half a sphere, so the area is given by $(4\pi r^2)/2$ = 12 x 49/2 = 294. Since one pot covers 12m^2, 24.5 pots are required to cover the whole dome once and 49 to cover it twice.

Question 35: B
If 49 x 2 = 98 litres of paint, therefore 98 x 0.4 = 39.21 litres solid volume when the paint dries, = 0.0392m^3. The volume of the hemisphere is $(4/3\pi r^3)/2$, dividing the reduction in volume by this gives 0.0392/686 = 0.0057%

Question 36: E
If the machine takes 400ms to wrap a sweet, it will wrap 10 sweets in 4 seconds = 0.4 seconds per sweet. Since there are 60 x 60 x 2 = 7200 seconds in 2 hours, and it can wrap 7200/0.4 = 18,000 sweets in 2 hours.

Question 37: C
If in 8 years time the sum of their ages is 52, the sum of their ages now is 52 minus two eights, which is 52 – 16 = 36. If they were both the same age, they would each be 36/2 = 18 years, however we are told John is 6 years older. Since the sum of their ages must still be 36, John has to be 21 and his brother is 15.

Question 38: B
The passage tells us that house prices have raised due to an increase in occupancy. Therefore making more houses available will make them more affordable. This could be achieved by either building more houses, or increasing the number of people in each house. Building more houses, C, is therefore a true conclusion based on information from right throughout the passage. The passage doesn't tell us what the current rate of house building is, it only speculates divorce as a cause of problems, and doesn't mention where divorcees live. Whilst it is true to say A, it is not the main conclusion of the passage, merely a paraphrasing of the first sentence, thus it is not the correct answer.

MOCK PAPER B SECTION 1 — ANSWERS

Question 39: C
If there are four stations between Crabtree and Eppingsworth, there are five independent journeys between them. Calculate the times between each of the stations, then sum them for the total journey time. The journey between stations 3 and 4 is leg 4 of the journey, and it takes 16 minutes. Therefore leg 5, the final component takes 16x0.8 = 12.8 minutes. Leg 3 takes 16/0.8 minutes = 20 minutes, leg 2 takes 20/0.8 = 25 minutes. Leg 1 takes 25/0.8 = 31.25 minutes. Summing these together gives:

31.25 + 25 + 20 + 16 + 12.8 = 105.05, about 105 minutes, since we do not measure rail journeys in fractions of minutes.

Question 40: E
The passage tells us residents fear the noise and disruption caused by the construction. Therefore E is the best answer, since it directly addresses the concerns the residents have. Residents would describe the other responses as unhelpful, since they do not address their main concern.

Question 41: D
To find out when the rental costs are equal, equate the two rental costs. If d is the number of days of hire, hiring from Tony's costs 23d, and hiring from Adam's costs 65 + 18d. Therefore at equality, 23d = 65 + 18d, 5d = 65, d = 13 days. However this is only at equality, and the question asks when a saving will be made by using Adam's. Therefore the answer is 14 days, when the first saving (of £5) is made by using Adam's.

Question 42: C
The passage tells us that antibiotic resistance could lead to people dying from Victorian diseases, and that liberal use of antibiotics in farming is the "most significant" contributor to this. Therefore it would be true to say that this use of antibiotics could cause serious harm.

Question 43: C
Only AMATAMA, option 2, is the same when viewed in a mirror and when viewed normally. The other options will not appear the same. To test this, you can use a shiny surface such as a mobile phone screen.

Question 44: B
You have to work through this sequentially. In week 2, 60% more purchases means 7,500 x 1.6 = 12,000 of purchases. A 30% profit margin means 1.3 x 12,000 = 15,600 of sales. In week 3, 2000 less in sales gives 13,600 in sales, and with a 60% profit margin, the purchase cost is 13,600/1.6 = £8,500.

Question 45: A
To answer this question, you need to work out the ratio of the areas. There are 6 squares and 8 triangles. If the area of a triangle is $3\sqrt{27}$ cm^2 = 0.5 x base x height, $6\sqrt{27}$ = the value of the base x height. Here, you might see that $\sqrt{27} = \sqrt{(36-9)}$, therefore $(\sqrt{27})^2 = 36 - 9 = 6^2 + 3^2$, and applying Pythagoras' theorem, the equilateral triangle side length works out at 6cm. Therefore this is also the square side length. As a result, the total area of the squares is $6 \times 6^2 = 216$. The total area of the triangles is $8 \times 3\sqrt{27}$. Therefore the total surface area is. $216 + 24\sqrt{27}$. The area of one square, which we are interested in, is $6^2 = 36$, so the probability of it landing on that portion of the area is $36/(216 + 24\sqrt{27})$. However this is not the simplest form. Factorise the denominator to $36/9(24 + \sqrt{27}) = 4/(24 + \sqrt{27})$.

Question 46: B
The passage tells us that the median length of commute is 40 minutes: therefore at least 50% have a commute which is equal to or shorter than 40 minutes, therefore B is true. Option E is not true, as the passage tells us that over 50% of *commuters* complain about the duration of their commute – but since not all people commute to work this might not be more than half of the population.

MOCK PAPER B SECTION 1   **ANSWERS**

Question 47: C
To solve this, calculate the number of possible outcomes ($3^3 = 27$) and the number of outcomes which contain no score. In every theoretical round, one third of outcomes have no score. Therefore after round one, there is a 1/3 = 9/27 chance of no score. In addition, in the next round there is a 1/3 chance of the other 2/3 being zero, and in the third round there is a 1/3 chance of the remaining 2/3 being zero. Therefore the probability of at least one zero score is (1/3) + (1/3 x 2/3) + (1/3 x 2/3 x 2/3) = 9/27 + 6/27 + 4/27 = 19/27.

Question 48: C
The graph shows that more water is required around the middle section for each additional increase in depth, with less at either end. Therefore the container is spherical.

Question 49: C
The passage states only problems with owning a listed building, and not a benefit. Therefore C is true – it is not a balanced argument, and there could be benefits to owning such a house that the passage makes no account of, therefore it might not be, on balance, bad to own such a property.

Question 50: B
The answer is found by an iterative process, multiplying each intermediate answer by the percentage decrease in a compound interest deduction. 36000 x 0.75 = 27,000. Keep taking ¾ off, repeating ten time to give the answer of 2,030 cars a decade later in 1015.

END OF SECTION

Mock Paper B: Section 2

1. *Design an experiment to deduce the sensitivity of a snake's hearing. Explain everything you would do, and your rationale for doing so.*

This question is different to most others you will encounter and does not demand the usual structure. Despite this, there should be a logical flow to the answer, explained comprehensibly and clearly, and reasoning should be justified adequately to provide support.

Introduction:

A brief overview of the independent and dependent variables would be appropriate, as well as a short hypothesis about what you may expect to observe.

Points to consider and justify:

- Independent variable = what you will change, probably the frequency or volume of the sound. You will be required to operationalize this variable, considering the equipment used to provide the sound – such as a computer, and the method of delivery to the snake – through speakers, through headphones etc.
- Dependent variable = what you will measure, to do with the snake's response to the sound. This could be operationalized in a number of ways, such as a detectable movement, which has been previously learnt by association, or simply a twitch – but this should be scientific and objective. Any equipment or materials, including an observer, should be identified.
- Sample and repeats = how many different snakes will you test? Will you repeat the experiment on the same snake a number of times?
- Setting = is this a laboratory experiment or will it be taking place in a naturalistic environment?
- Control variables = it must be clear that you have considered potential confounding factors and have taken steps to control them. These may be extraneous noise (control for by taking place in a soundproof room), errors in determining when the snake has heard the noise (repeats are probably the easiest way of eliminating this), sample bias – are you testing just one breed of snake or lots of different ones? Observer bias – if an observer is detecting movement, ensure they are doing this reliably.
- Interpretation of results – include how you will present your data and suggest any data manipulation techniques (graphs, statistical tests) outlining what you hope this will achieve.
- Reporting the experiment – explain how you might wish to write up the experiment and any further research you may want to highlight.

Ideas for justification:

- Convenience
- Time taken
- Accuracy
- Validity
- Reliability
- Access/understanding

Conclusion:

Summarise the experiment in a couple of sentences AND your overall justification about what you wish to achieve, identifying both the positives and potential limitations or need for further study.

2. "The eternal mystery of this world is its comprehensibility"

To what extent is the world comprehensible?

Introduction:

- Include a definition of comprehensible – understandable, intelligible
- You may wish to consider the extent to which comprehensibility in itself is subjective, and therefore your essay will be biased towards your own views on comprehensibility.
- Introduce a summary of the ideas you want to explore within the essay on both sides – arguments for the world being comprehensible and arguments against (see ideas below)

Points for comprehensibility:

- We are restrained by the constraints of our own human mind – is the question how much can the human mind comprehend? The world is limitless until we reach the limits of our own minds.
- We are still discovering more and more about the world and haven't reached our limit yet – there is always more to discover and further our knowledge and understanding. Until we can no longer advance in our scientific understanding, the world is comprehensible and in our hands to discover.
- One could argue that we are the most advanced and evolved organisms in the world and therefore everything we need to understand must be less complicated than ourselves.
- With an understanding of history, we can understand why things are the way they are – as long as we learn from the past, surely there is no more in the future we cannot reach?

Points against comprehensibility:

- One could argue that the world is contained within our minds, and since we are within our minds we cannot objectively understand them – do we have to be outside of the world to comprehend it, and therefore not be of this world?
- The human mind may be so complex that it is, paradoxically not complex enough to understand itself.
- There must be a limit to our understanding, since the world has limits and our brain is a limited place for processing.
- We do not know what else there is to comprehend, and therefore we will never know when we have reached the limits of our comprehensibility – therefore never fully reaching complete understanding.
- Would we be God if we could comprehend the world and everything in it?

Conclusion:

- Include a summary of all points made, and give a balanced overview of both sides of the argument.
- Reference back to the original quote – it may be relevant to emphasise the word 'mystery' in the context.
- Draw together your points in some kind of concluding statement – either to say the world is completely comprehensible or to say that it is only comprehensible to some extent would be justified, so long as this is logically and critically summarised and justified.

3. "The greatest obstacle to learning is education"

Argue for or against this statement.

Introduction:

- Consider a definition for learning – the acquisition of knowledge; versus education – a structured approach to transferring information about the world from one person to the other.
- Identify the ways through which knowledge can be acquired – through teaching, experience, study – emphasise that learning doesn't have to be purposeful, whereas education involves the intention of learning (whether or not it is achieved).
- This essay asks for ONE perspective – introduce the argument you wish to pursue.

Possible arguments for:

- The greatest feats of learning could be considered the acquisition of language and/or movement – all of which are done without and before education.
- Education is directive – based on the views of a few people about what they believe is important – and therefore it narrows our minds to only the concepts taught.
- Everyone learns in different ways – creatively, actively, through vision, through hearing and education, as we know it is too structured to suit everyone.
- School is a bubble – set apart from the real world – you could argue that the real learning happens once children leave school and have to fend for themselves.
- Have homeless people or those with no access to education learnt less? They may have learnt different skills, but broader, more practical ones.
- Learning involves actively engaging with the information given, and requires an understanding – you could say that education teaches simply recall of facts, not true knowledge of the world.
- Examples of people who have achieved much without a formal education e.g. Alan Sugar

Possible arguments against:

- You could say that education is feeding the natural curiosity within human beings – we naturally want to learn more about the world and formally educating people of this is an easy and natural way to pass on knowledge.
- Education provides not just academic knowledge, but social knowledge; how to behave, what is right and what is wrong, how to make friends. There isn't such a concentrated opportunity to learn these things in other places.
- Reading and writing are basic skills required for so many careers – education opens the doors to further learning
- How could one say that a person has learnt nothing after graduating from school? Therefore is cannot be a barrier to learning, because children surely learn something!
- Knowledge needs to be passed on to those capable of furthering it – and therefore education provides a means of passing on knowledge and developing our understanding of concepts – if education didn't exist, development would be slower so we must be learning something.
- Uneducated people make up the majority of our unemployment figures and are often more likely to turn to drugs, alcohol and develop other social problems.

Conclusion

This is an argumentative essay so the conclusion MUST reach a decision. Summarise the key ideas of the essay and dismiss any opposing perspectives.

4. Does a vacuum really exist?

Introduction:

- This must include a definition of vacuum – a space entirely devoid of anything.
- You may wish to clearly indicate what you mean by 'anything' – matter, time, etc.
- This could be considered philosophically or physically – make it clear which (or both) perspective you are wishing to take.
- Introduce the key ideas on both sides you wish to pursue on both sides of the argument – reasons for it existing and reasons against (see below)

Arguments for:

- How can we prove that 'nothing' exists – in proving there is nothing, there must be something, in order for there to be something to prove.
- Just because our brains cannot comprehend a place where there is nothing, that does not mean that it doesn't exist – we just may not be adequately equipped to understand it.
- We can never prove that a vacuum doesn't exist, because there is no way of finding it if there is nothing there – so we must assume it exists if we cannot disprove it.
- If there is a place where there is something, there must also be a place where there is nothing, in order for the place where there is something to be valid.

Arguments against:

- Without matter, there is an absence of anything, and therefore without anything, there is nothing – so it cannot exist.
- There are fields and properties and relativity everywhere in the known universe, provided the physics applies equally everywhere, which means there is always something there, so there is nowhere in the universe that there is nothing.
- In order for something to exist, it must have properties which can be defined and demonstrated within the realms of our reasoning – since this is not the case for vacuums, they may not exist.
- Our understanding of physics is by no means complete and therefore our ideas about what matter is and how to define it are likely to change – therefore there is no reason to believe there must be a place where there is nothing – it could just be filled by something we haven't discovered yet.

Conclusion:

- Draw together all the points made on both sides of the argument – summarise the key points and bring them down to earth again.
- The conclusion must be solid and clear – even if not particularly complicated, it must be logical and understandable, in order to bring together this potentially confusing subject.
- You may wish to reach a decision, or decide that it depends on the perspective you take – as long as this is correctly justified.

END OF PAPER

Mock Paper C: Section 1

Question 1: D
Answer C) is completely irrelevant, so is not a flaw. Answer B) is not a flaw because when assessing an argument, anything that is stated (i.e. not concluded from other reasons in the passage) is accepted as true. We do not require evidence or sources for any statistics presented. Answers A) and E) are both claiming that something is immoral, which is thus expressing an opinion on the part of the arguer. This is not a flaw, the arguer is at liberty to claim something is immoral, and to claim that the government is morally obliged to act, and that it has not done so. However, answer D) identifies a valid flaw. The argument rests on us accepting that if there were less uninsured drivers, there would be less crashes. This is not necessarily correct, so D) is a flaw in the passage.

Question 2: A
The passage does not say anything about whether Brazil should legalise guns, it simply reports what one commentator *said* was the reason why the move to ban guns was unsuccessful. Thus, C) and D) are not best supported or opposed by the passage as directly as A) and B). The passage clearly indicates that the UK *should not* legalise guns, when it says "legalising ownership in the UK would be a bad move". Thus, A) is best supported by the passage. Since A) is best supported, E) is also incorrect, and the answer is A).

Question 3: C
For each of the walls where there is no door, the wall is 6 tiles high and 5 tiles wide, which is 30 tiles. The wall where the door is requires a row of 2 tiles above the door, then there is a width of wall of 120cm which requires completely tiling, which is 6 tiles high and 3 tiles wide, hence this wall requires a total of 20 tiles. Hence a total of 110 tiles are required for the walls. The floor is 2 metres by 2 metres, so 5 tiles by 5 tiles, hence 25 tiles are required for the floor. Hence the answer is 135.

Question 4: D
Answer E) is irrelevant to which of Trevor and Jane will arrive first, so does not weaken the conclusion. Answers A), B) and C) all strengthen the answer, giving further reasons why we might expect Trevor to arrive first. Answer D), however, would slow Trevor down, meaning that it was more likely that Jane would arrive first. Thus, Answer D) weakens the passage's conclusion, and hence Answer D) is the answer.

Question 5: A
He has enough butter to make 2.5 times as many cupcakes as the recipe, which is 50
He has enough sugar to make 3 times as many cupcakes as the recipe, which is 60
He has enough flour to make 5 times as many cupcakes as the recipe, which is 100
He has enough eggs to make 3 times as many cupcakes as the recipe, which is 60
The lowest of these is 50, so he can make 50 cupcakes. He needs 2.5 x 4 eggs to do this, which is 10 eggs. Therefore he has 2 eggs left over.

Question 6: B
Let the number of minutes that the journey takes be 't'. So ABC charge 400+15t pence for the journey. We can calculate that XYZ taxis charge 400+(30x6) = 580 pence. In order for both journeys to cost the same, 580=400+15t. 180=15t, thus t=12. Therefore the 6 mile journey needs to take 12 minutes. 6 miles in 12 minutes is 30 miles per hour, so the answer is B.

Question 7: E
We can see that all of answers A) through D) are essential for the conclusion to be valid from the squire's reasoning. Lancelot must have great courage, this must be a requirement for the Adzol, and no other knights must have sufficient courage, in order for us to be certain that Lancelot will succeed but all of Arthur's other knights will fail. Thus A) and B) can be clearly identified as assumptions. C) and D) require a bit more thought, but we can see that nothing in the passage explicitly states the Elders' tales are correct. If the elders are not correct, then great courage may not be required to be successful in the Adzol. Thus, both C) and D) are also assumptions. Hence, the answer is E).

Question 8: E
B) is incorrect, as the passage does not say that arch-shaped gaps *always* indicate where windows once stood, simply that *these arches* do. C) is also incorrect, as the passage simply states that windows are not found in *underground halls*. A) is a reason in the passage, and is not a conclusion. D) and E) could both be described as conclusions from this passage, but we see that if we accept D) as true (along with the fact that the hall is now underground), we have good reason to believe that E) is true, whereas E) being true does not necessarily mean that D) is true. Thus, E) is the *main* conclusion, whilst D) is an *intermediate conclusion*, which supports the main conclusion.

Question 9: D
Usually bread rolls cost 30p for a pack, but if the cost per bread roll is reduced by 1p then a pack will cost 24p. Hence we need to find z, where $24(z+1)=30z$, where z is the original number of packs that could have been afforded. $24z+24=30z$, hence $24=6z$, so $z=4$. Hence he was originally supposed to be buying 24 bread rolls.

Question 10: B
Answer E) is an irrelevant statement that says nothing about whether England *do* have good players. Answers A) and D) actually weaken the sporting director's arguments, suggesting that England may have a good team, and it may just be poor performances in world cups, and not a lack of talented players. This leaves B) and C). C) may appear to strengthen the sporting director's argument, but on closer inspection we see that in fact it says that for the last 70 years, England have had at least 1 player in the top 10 in the world. This does *not* strengthen the argument that England have been lacking talent for the last 25 years, and may actually reinforce the chairman's argument that it is simply the *current* crop of players that are not good enough. Answer B), however, does strengthen the argument, suggesting that England's performances have been poor over the last 20 years, thus strengthening the argument that there may be a lack of talented players that has been ongoing for a couple of decades, as claimed by the sporting director.

Question 11: A
He can prepare each batch of cakes while the previous one is in the oven but it takes longer so we have to allow 25 minutes for each batch, plus 20 minutes for the last batch to cook while no further batch is being prepared. There are 12 in each batch, so for 100 cupcakes there needs to be 9 batches. Hence the total time needed is 25 minutes x 9, + 20 minutes. This is 245 minutes, or 4 hours 5 minutes. Hence to be ready by 4pm he needs to start at 11:55am, so the answer is A.

Question 12: D
We can first work out the rate of girls' absenteeism. First we need to work out how many of the pupils at Heather Park Academy and Holland Wood Comprehensive are girls. Let g be the number of girls in Heather Park Academy. Then $0.06(g)+0.05(1000-g)=(1000)(0.056)$. Then $0.06g-0.05g=56-50$. Then $0.01g=6$, so $g = 600$. Hence 600 pupils at Heather Park Academy are girls. The proportions at Holland Wood Comprehensive are the same but there are half as many pupils, so 900 pupils at the two schools combined are girls.

The average absenteeism of girls is 6.1%. We know that 900 of the 1100 girls have an average absenteeism rate of 6%. Let the average absenteeism rate of girls at Hurlington Academy be r. Then $900 \times 0.06 +200r = 0.07 \times 1100$. Hence $54+200r=77$. $77-54 = 200r$. $23/200 = r$. $r=0.115$. Hence, the rate of absenteeism amongst girls at Hurlington Academy is 11.5%

Question 13: C
A), B) D) and E) are all directly stated in the passage, so can all be reliably concluded. Perhaps the trickiest of these to see is answer D), which is true because the passage says "*due to*" the advent of more accurate technology, thus clearly identifying that this had *caused* the switch to the situation of most watches being made by machine. C), however, is *not* necessarily true. The passage states that most *watches* are produced by machines, but only states that *some* watchmakers now only perform repairs. This does not necessarily mean that most watchmakers do not produce watches. It could be that only a handful are required in the entirety of the watch industry for repairs, and that the numbers still producing watches exceeds those in the repair business. Thus, C) cannot be reliably concluded from the passage.

MOCK PAPER C SECTION 1 — ANSWERS

Question 14: A
B) is not a valid conclusion from the passage, because the fact that someone uses an illogical argument (as some Pescatarians are claimed to in this passage) does not mean that they cannot use logic. D) and E) are not conclusions from this passage because the passage is not saying anything about the ethicality of eating meat, but simply commenting that one argument used against doing so is not logical. Answers C) and A) are both valid conclusions from the passage, but we see that if we accept C) as being true, it gives us good cause to believe that A) is true, but this does not apply the other way round. Thus, C) is an intermediary conclusion, whilst A) is the main conclusion.

Question 15: E
The research conducted does not ask about whether it is *important* to learn some of the language before travelling abroad, simply whether participants *would*, so B) cannot be concluded. D) is incorrect because the passage states *15%* would, which is clearly not less than 10%. The passage states that this is symptomatic of a deeper underlying issue, but does not say that many issues of racism stem from this, so C) cannot be concluded. Now, the passage states that 60% of people feel foreign people should learn English before travelling to Britain, and 15% of people would attempt to learn the language before travelling to a country which did not speak English. However, this 15% could be some of the same people as the 60%, in which case A) would be incorrect. Thus, A) cannot be reliably concluded. However, there must be at least 45% of people who feel that foreign people should learn English, but would not learn a foreign language themselves, so E) *can* be reliably concluded.

Question 16: D
She needs to print 400 x 2 = 800 double sided A4 sheets, which will cost £0.03 each. Hence the total cost of this is £240. She also needs to print 1500 single sided A5 sheets, which will cost in total £150. Hence the total cost is £390.

Question 17: B
The passage has stated that if Kirkleatham win the game they will win the league, so E) is not an assumption. Meanwhile, the manager has stated A), C) and D), and the passage has not claimed anything about whether Kirkleatham can easily win the game, so A) and D) are not assumptions. However, B) does identify an assumption in the passage. The fact that Kirkleatham will not win the game without playing with desire and commitment does *not* mean that they will win the game if they do play with desire and commitment. And we can see that for the argument's conclusion (that Kirkleatham *will* definitely win the league) to be valid from its reasoning, this is required to be true. Thus, B) identifies an assumption in the passage.

Question 18: B
Answers A) and E) are not relevant, because neither affect the strength of the councillor's argument from a critical thinking point of view. The councillor's argument says nothing about house prices, simply the cost of building the estate and the effects on Wildlife, so A) is not relevant. E) is not relevant because additional support, or likelihood that it will be heeded, does nothing to affect the strength of a given argument. C) and D) actually strengthen the councillor's argument, suggesting that brownfield land does have good infrastructure (C)) and that the greenbelt areas do have a lot of wildlife (D)). B) does weaken the councillor's argument, as it suggests that building on brownfield land may also have adverse impacts on wildlife.

Question 19: B
The reasoning in the passage can be summarised as "IF A happens, B WILL happen. IF B happens, C WILL happen. Thus, if A happens, C will happen". Only B) also follows this reasoning. A) can be summarised as "If A does NOT happen, B will happen. A doesn't happen, so B will happen". C) can be summarised as "If A happens, B will happen. Thus, if A happens, C can happen". Here, there is no explanation of "If B happens, C will happen", so it is not the same reasoning as in the question. D) reasons as "If A happens, B will happen. If A doesn't happen, C will have to happen. Thus, to prevent B happening, C will have to happen", which is vastly different to that in the question. E) is the direct opposite of the question, reasoning as "If A doesn't happen, B can't happen. If B doesn't happen, C can't happen. Thus, if A doesn't happen, C can't happen".

MOCK PAPER C SECTION 1 — ANSWERS

Question 20: B
We can tell the amounts for the green party and the blue party are both 1/3 of the total, and that the amount for the red party is 1/4 of the total. Hence 1/12 is left, so the amount for the yellow party must be 1/12. Hence the red party have 3 times the intended vote of the yellow party.

Question 21: B
A large pizza with mushrooms and ham is £12, garlic bread is £3, chips are £1.50 x 2 = £3, a dip is £1, hence the current total is £19. The cheapest way to order this is to get the price up to exactly £30 as this will reduce the price to £18. This takes £11. Only one of these options costs £11, which is a large pizza with mushroom. Hence the answer is B.

Question 22: D
The reasoning in the passage follows the style: "If A doesn't happen, B can't happen. If B doesn't happen, C will happen. Thus, if A doesn't happen, C will happen". Only answer D) follows this style. Answer A) simply reasons as "If A happens, B will happen. A doesn't happen, so B won't happen", which is both incorrect and a different style from the question. Answer C) is also incorrect, reasoning as "A is being considered in order to do B. If B happens, C will happen. However, A does not happen, so C won't happen", again both incorrect and a different style to the question. Answer E) follows the pattern "If A happens, B won't happen. If B doesn't happen, C will happen. A happens, so C will happen". Answer B) is similar to answer D), but not quite the same. Answer B) follows the pattern of "If A does not happen, B will not happen. If B can't happen, C can't happen. Thus, if A doesn't happen, C can't happen". In answer B) and the question itself, C *will* happen if A doesn't happen, so there is a slight difference in Answer B). Hence, the answer is D).

Question 23: D
The question simply describes how a combination of factors are responsible for the M1 Abrams being the world's most formidable tank, so the view of country X is incorrect. It does *not* claim that it is impossible for a tank to be as good as the M1 Abrams, so E) is not a valid conclusion. Equally, it does not say the new tank's armour will not be as good as the Abrams (in fact it is implied that it may well be as good), so C) is also incorrect. A) B) and D) are all valid conclusions from this passage, but we can see that A) and B) contribute towards supporting the conclusion in D). Thus, D) is the main conclusion of this passage, whereas A) and B) are *intermediate* conclusions given to support this main conclusion.

Question 24: D
We can simply add up the amounts in the bank accounts and find the difference between each month – it doesn't matter that the salary is paid in as it is the same every month. Doing this, we find out the biggest difference is between 1st May and 1st June, hence the answer is D, May.

Question 25: B
Firstly we can work out the full dose for the son. He needs to take 0.2ml per kg of weight for each dose, and he is 40kg, so this is 8ml. He takes 40 doses altogether, so in total he needs 320ml of medicine.
Then we work out the doses for everyone else, add them together and halve them. The daughter's full dose would be 2ml, 30 times, which is 60ml altogether. My dose would be 7.5ml, 60 times, which is 450ml. My husband's dose would be 8ml, 60 times, which is 480ml. Altogether, this is 990ml. However, only half of these dosages is needed, which is 495ml. Hence the total needed is 320 + 495, which is 815ml. Hence 5 200ml bottles of medicine are needed for the full course.

Question 26: A
There are 21 forks and 21 knives. If half as many are red as blue, and half as many are blue as yellow, they are in the ratio red:blue:yellow 1:2:4. Hence of the 21, 3 are red, 6 are blue and 12 are yellow. Hence the probability of getting a yellow knife is 12/21 = 4/7. The probability of getting a red fork is 3/21 = 1/7. Hence the probability of getting both is (1/7) x (4/7) = 4/49.

MOCK PAPER C SECTION 1 — ANSWERS

Question 27: D
The Principle used in the passage is that public funds raised through taxation (which is compulsory) should not be used for any services unless they benefit everyone, such that nobody is forced to pay for services that do not benefit them. Answer D) is the best application of this principle, as it directly follows it. Answer B) mentions public funds being used to support a service that benefits the whole country, but this does not necessarily mean that they *shouldn't* be used to support services that don't benefit everyone, so answer B) is not as directly an application of the principle as answer D). Answer E) is not the same principle because this is talking about funds being used for services that benefit the *country*, rather than everyone in it. Meanwhile, Answer A) is talking about how many people *use* a certain service, rather than how many people *benefit* from it, so this is not the same principle. Answer C) is completely different, talking about funds being used because some cannot afford private health service, regardless of how many people are benefitting from the public health service.

Question 28: B
The chairman has *stated* that 'All Inclusive' services are more popular than 'Hourly' services. He has not deduced this from any evidence, and thus he has assumed nothing about their popularity. Thus, C) and D) are incorrect. The chairman's argument is simply that focusing on 'All Inclusive' services will bring in more profit than 'Hourly' services, as he says they should focus on 'All Inclusive' *rather* than 'Hourly' services. Thus, any reference to other services or other profit-raising strategies are irrelevant, so A) and E) are irrelevant. However, B) correctly identifies the chairman's flaw. Just because 'All Inclusive' is more *popular* than 'Hourly' services does not mean they are more *profitable*, and if they are not then the chairman's conclusion is no longer valid. Thus, B) correctly identifies the flaw in his argument.

Question 29: A
B) is not an assumption because the passage *states* that renewable sources do not cause damage, so we accept this as true. E) is not a flaw, because again the passage has stated that the use of these fuels to produce power will continue to cause climate change *as long as it continues*, thus we must accept as true that it cannot be halted or prevented whilst these fuels are used. C) and D) are irrelevant to the argument's conclusion that if we wish to stop damage to the environment, we need to switch to renewable fuels, and thus they are not flaws. However, at no point is it stated that *all* non-renewable fuel sources cause environmental damage, it is only stated the non-renewables *such as* Oil, Coal and Natural Gas do.
Thus, we have no guarantee that Nuclear fuel will cause environmental damage, and if it doesn't, the passage's conclusion no longer stands. Thus, A) is a valid assumption in the passage.

Question 30: E
Answers A) and D) do *not* strengthen or weaken the argument because the question states that this increase in non-vaccinated individuals has occurred despite powerful evidence of vaccine safety, and in spite of advice from doctors. This suggests that people who do not vaccinate pay little attention to evidence of advice from doctors, so we should not expect these factors to have much of an effect. B) is completely irrelevant to whether the rate of increase will continue. C) actually weakens the argument, suggesting that such increases are common, and normally stop after 6 years. If the current increase was to follow suit, it would stop next year and vaccination rates would not fall below 90%. E), however, implies that this kind of increase has happened only once before, and in this case it continued for 13 years. If the current increase was to follow *this* pattern, it would continue for another 8 years, where vaccination rates would be below 90%. Thus, E) strengthens the argument's conclusion that we should expect an outbreak of measles.

Question 31: A
Answers B) and E) are both in contradiction with stated points in the question, which states that there is now powerful evidence human bodies *are* set up for long-distance running, and that it is well established that humans evolved in Africa. Answer C) is irrelevant, because the presence of other theories does *not* necessarily affect whether we should believe this theory based on the new evidence. Answer D) actually strengthens the argument, suggesting that the new evidence does provide powerful reasons to believe this theory. Answer A) however is a valid flaw, that evidence supporting a theory does not necessarily *prove* that it is true. Thus, the answer is A).

Question 32: C
The percentage of students who had their grades predicted correctly is the same as the number who had their grades predicted correctly as there are 100. Hence we simply need to add up the numbers on the diagonal of the table, where actual grade is the same as predicted. This adds up to 39, hence the answer is C.

Question 33: B
The 2 wage reductions mean that when the wage increases happen, the raises will be x% of a smaller number than the decreases were. Thus, the wage will not rise as high as the original level.
If you are struggling to visualise this, the easiest way to do it is to substitute a number for x. Let us do the calculation treating x as 10.
The first wage drop is by 10%. Thus, the wage is now 90% of the original wage.
The second wage drop is also by 10%, but at this point, the wage is only 90% of the original wage. Thus, the drop will be by 10% *of 90%* of the original wage, resulting in a new wage of 81% of the original wage (10% of 90% is 9%)
Then, we have the first increase, which will be 10% of this new wage (81% of the original wage). Thus, after the first increase, the new wage will be 81% +(10% of 81%) of the original wage. Thus, it will be 81% + 8.1%, which is 89.1% of the original wage.
Now we have the second increase. Another 10% is added, this time of 89.1% of the original wage. We now have an increase of 8.91% (10% of 89.1). Thus, after the final raise, the wage will be 89.1% + 8.91%, which is 98.01% of the original wage. Thus, the new wage is lower than the original wage.

Question 34: C
The passage says nothing about whether it is more important to have good doctor-patient relations than scientific progress, so answer E) is not a valid conclusion. Answer D) is a direct argument against the question, which claims the naming system *should* be changed, so answer D) is not a conclusion. The passage states that the confusing system causes problems in scientific literature, but this does *not* necessarily mean that changing it would allow faster progress in scientific research, so answer B) is not a valid conclusion. Answers A) and C) are both valid points from the passage, but we can see that Answer A) is a *reason* stated in the passage, which helps support the statement in C). Thus, Answer C) is the main conclusion.

Question 35: E
180 people applied in total. 128 did French, 64 did Spanish, 48 did neither French nor Spanish. We hence know that exactly 132 people did either French, Spanish or both. Since 128 people did French, only 4 people can have done Spanish without doing French. Hence 60 people must have done both French and Spanish.

Question 36: E
Answer A) directly strengthens the argument, suggesting that increased funding on education cannot cause a larger reduction in drug usage. Answers B) and D) do not directly affect the argument either way. Both these answers concern whether providing youth centres/recreational activities in areas of high social deprivation can cause reductions in drug usage. However, this does not directly affect the argument's conclusion that education alone cannot provide a further reduction in drug usage. Answer C) also does not affect the answer, for similar reasons, as it does not necessarily affect whether education can or cannot provide a further reduction in drug usage. Answer E), however, directly weakens the argument, suggesting that Spain has achieved a greater reduction in drug usage than the UK, simply by providing education. This directly challenges the argument's conclusion that education alone cannot further reduce drug usage.

MOCK PAPER C SECTION 1 — ANSWERS

Question 37: D
During June and July there are 61 days. The show will have to travel 8 times in the two months, taking 2 days each time, so there can be a maximum of 61 – 16 = 45 days of performances. Given the musical spends the same number of days in each place, this means 5 days of performances in each location. Given there are 5 days of performances and 2 days of travelling, which totals a whole week, the show must always travel on the same days of the week and perform on the same 5 days. The best time of the week to travel is Thursday and Friday because this means missing no matinee performances, so the most performances happen when each location's performances start on a Saturday and finish on a Wednesday. This means there are 8 performances in each location, a total of 72 performances. Hence the maximum number of people that can see the show is 72,000.

Question 38: A
If every swimmer swims 3 identical times, then each swimmer will have 3 consecutive ranks. Hence the best swimmer will have swims 1, 2 and 3, the second will have 4, 5 and 6, etc. Hence the first swimmer's total score is 6, the second's is 15, the third's 24 (7+8+9), the fourth's 33 (10+11+12), the fifth's 42 (13+14+15), the sixth's 51 (16+17+18), the seventh's 60 (19+20+21), the eighth's 69 (22+23+24), the ninth's 78 (25+26+27), the tenth's 87 (28+29+30). Hence 9 swimmers proceed to the final as the tenth best swimmer has a score of over 80.

Question 39: B
The question states A), B), C) and D), so all of these can be reliably concluded. However, the passage does *not* state B). The passage claims that *most* of the public in Europe is scared of nuclear power, and that some point to the disasters mentioned as evidence that it is not safe. However, this does not necessarily mean that these disasters have caused these fears. It could be a minority using these as evidence, and that the rest of the public has other reasons for fearing nuclear power.

Question 40: C
If Megan's mixture contains 3 times as much water as squash, then it must contain 9 litres of water and 3 litres of squash. If bottles of squash contain 1 litre to the nearest decilitre then they can contain as little as 950ml. Hence to guarantee she has 3 litres of squash, Megan must buy 4 bottles of squash.

Question 41: A
Andrew runs 3/4 of the distance that Alice does in the same time, therefore he must run at 3/4, or 75%, of the speed that Alice does. Amanda runs at 60% of the speed that Alice does. Hence Amanda runs at 60/75, 4/5 of the speed that Andrew does. Putting it a different way, Andrew runs 1.25 times quicker than Amanda. If Amanda runs at 200m per minute, then Andrew will run at 200 x 1.25 = 250m per minute. 10000/250 = 40, hence Andrew takes 40 minutes to run the race.

Question 42: E
Although it comes at the beginning of the passage, E) is actually the main conclusion. We can see that if we accept B), C) and D) as being true, along with the fact that the NHS needs more funds to deal with the extra workload caused by the ageing population, then we have good reason to believe E). Thus, E) is the conclusion, and B), C) and D) are all reasons supporting it. A), meanwhile, is a counter argument, which is then refuted in the passage.

Question 43: A
England won Pool C so they will be in Quarterfinal 3, where they will play Brazil. If they win, they will play the winner of Quarterfinal 1. Hence they can only meet teams from Quarterfinals 2 or 4 in the final. These teams are Argentina, Nigeria, South Africa or Holland. Hence the only one of these 5 teams they can play in the final is Nigeria.

Question 44: D
To make 1 metre she needs 4 of the 300mm length or 5 of the 210mm length. There are hence 3 joins of length 210x5 = 1050mm, totalling 3150mm, and 4 joins of length 300x4 = 1200mm, totalling 4800mm, on each side. Hence on each side there is 5850mm of tape. Hence the total tape is 10,700mm or 10.7 metres.

Question 45: E
The team must have played 12 matches at home and 12 away. If they have won 18 matches altogether and have won twice as many matches at home as away then they must have won 12 matches at home and won 6 matches away. They have hence lost no matches at home and 6 matches away. Therefore the answer is 6.

Question 46: C
The passage has *stated* that the fear is illogical, so E) is not a valid assumption. Meanwhile, whether there are other contributing factors is irrelevant to whether sci-fi has caused some cases of public fear of scientific progress, and the passage has not assumed it is the sole reason, so D) is not a valid assumption. B) actually supports the passage by reinforcing one of its reasons (how the viruses in many sci-fi movies are impossible), whilst A) is not a flaw because the passage states how *many* sci-fi movies are responsible for causing illogical fear of science. However, the passage has assumed that deaths could be prevented without the hindrance to progress from sci-fi movies, and this does not necessarily follow on from its reasoning. Thus, C) is a valid assumption from this passage, and C) is therefore the answer.

Question 47: B
Ashley walks 8km at 8km an hour, therefore he takes 1 hour to get there, so he leaves at 12pm.
Ben gets the bus, which takes 40 minutes and departs at 25 minutes past or 55 minutes past. If he gets the 12:25 bus he will be late so he needs to leave at 11:55am.
Callum cycles 12.5km at 12km an hour, which will take him 62.5 minutes, so he needs to leave at 11:57:30am.
Dave gets the train which takes 20 minutes and goes every 10 minutes, so the earliest he will need to leave is 12:20pm.
Ed gets the park and ride bus which comes every 10 minutes and takes 15 minutes, plus a 10 minute drive, so the earliest he will need to leave is 12:25pm.
Hence Ben needs to leave the earliest.

Question 48: B
If the 6 friends all send each other cards, this is a total of 30 cards. 5 of these are to Sophie and 5 are from Sophie. Hence 20 are priced at £0.50, which is a total of £10. A further 5 are priced at £1.50, which is £7.50. The final 5 are priced at £1.20, which is a total of £6.00. Hence the total cost of sending the cards is £23.50.

Question 49: E
In Rovers' first 3 games, they have scored 1 goal and had 8 goals scored against them. In total they scored 1 goal and had 10 goals scored against them, so they must have lost their last game against United 2-0.
In City's first 3 games, they scored 7 goals and had 3 goals scored against them. In total they scored 10 goals and had 4 goals scored against them. Hence they must have won their game against United 3-1. Hence the answer is E.

Question 50: D
Answer E) is not a valid conclusion from the passage, which makes no reference to whether examinations are the fairest method of assessment, simply that they are the fairer than coursework. The passage makes no reference to whether the education minister is logical, so A) is clearly incorrect. B), C) and D) are all valid conclusions from the passage, but we can see that B) and C) contribute towards supporting answer D). Thus, we can see that D) is the main conclusion of this passage, whilst B) and C) are intermediate conclusions, which support this main conclusion.

END OF SECTION

Mock Paper C: Section 2

1. <u>To what extent are 'logical' and 'rational' synonymous?</u>

Introduction:

- Consider a definition or overview of each of the terms; rational meaning behaviour or arguments, which are in accordance with reason; logical being in accordance with the rules of formal argument to produce sound inferences.
- Perhaps give some examples of logical and rational situations that illustrate your definitions.
- Introduce a summary of your arguments both for and against the two being synonymous.

Arguments for synonymous:

- Both words are often used interchangeably – a sound explanation can be described as 'logical' and 'rational'.
- We need to be rational to be able to reason logically – logic requires rationality.
- When looking for sensible solutions to situations, adopting a strategy of logic is often an effective way to problem solve and process information.
- Behaviour is often described to be irrational if it doesn't appear to be a logical response to a situation e.g. the logical place to go if you have no money in your purse is the bank, not a shop.
- Irrational behaviour often occurs when people are unable to take time to think through or process situations (e.g. due to emotion or time pressure), and thus do not implement logical reasoning.
- Logic is an effective way of modelling decision making in computer programming.
- Consider the consequences of a world without logic – chaotic, no structure, inability to make the correct decisions in situations.

Arguments against synonymous:

- Rational decisions can be made without fully processing every situation and considering the logical inferences based on each premise.
- Just because behaviour is illogical, does it mean it is irrational?
- Rational responses to certain situations may consider other factors such as emotion, the influences on other people, long term consequences. e.g. it is totally rational to want to avoid a situation that may be the logical decision to make if it will cause personal distress.
- Logic may be too simple to explain all situations.
- Some situations may be unable to be modelled logically or there may be no one logical outcome, yet rationality is still possible.
- Just because there is no logic without rationality, doesn't mean there is no rationality without logic.
- Examples of situations where logic fails or when a logical solution is not possible.

Conclusion:

- Include a summary of arguments both for and against synonymy and draw together both sides of the argument.
- Come to some form of a conclusion – probably a middle ground between logic being one, but not the only, possible way of modelling rationality.
- Emphasise the limitations of simplifying rationality to only logical explanations.

2. *In what instances is aggression justified?*

Introduction:

- Include a summary or definition of aggression – behaviour implemented with the goal of instigating harm to other people.
- Give examples of different types of violence and indicate what types of violence you will be considering in your essay – physical, emotional.
- You may wish to make it clear from the beginning where you stand with regards to the argument – whether it is ever justified, in certain situations or you may wish to suggest here that it is never justified.
- However the essay is also looking for a balanced argument – so summarise both sides, suggestions of when it may be considered justified, but also arguments why this may not be the case.

Possible examples of justified situations:

- When protecting oneself from a physical attack.
- When protecting a loved one from a physical attack.
- The same as above, but verbally- you may wish to consider these two scenarios separately.
- When standing up for a cause e.g. violent campaigns.
- War – fighting for your country.
- In order to control a criminal e.g. police violence towards criminals (may be linked to protection).
- When the target is not another living thing e.g. at the gym, or to let out frustration towards another person on an inanimate object.
- When the target agrees to/or asks for aggression to be displayed – e.g. 'hit me if I do…'
- In jest, comedy, acting e.g. play-fighting.
- In sport e.g. boxing, rugby.

Possible reasons against justification:

- We should never allow aggression in any situation since it makes it more socially acceptable as a behaviour in other, non justified situations.
- Aggressive responses are often learnt from other people therefore we should not model them, in any situation even sport or comedy.
- There are always other ways of letting out frustration, which may be more effective or adaptive, than being aggressive.
- Even if the immediate consequence is not negative, long-term implications of exposure to aggression may not be positive.
- Aggression may also be harmful to the aggressor, as well as the target.
- Any form of aggression has the potential to escalate to cause danger.
- You should never 'fight fire with fire' - Aggression by one person encourages aggression in another.
- Religious/ethical reasons e.g. 'turn the other cheek'.

Conclusion:

- Summarise reasons when aggression is generally not justified.
- Counteract these with a few examples or generalised situations.
- Argue against this to some extent to summarise the other side of the argument.
- You may wish to reach a conclusion about whether aggression is ever justified, and if so, a generalised description of when.

MOCK PAPER C SECTION 2 — ANSWERS

3. *What are the limitations of scientific theories of human behaviour?*

Introduction:

- Explain what a scientific theory is - a generalised, well substantiated explanation of situations or natural occurrences, used as a model and to predict future events or occurrences
- Explain how this may be achieved – 'scientific method' usually requires observable, experimental, objective evidence, repeatedly tested and experimented across multiple situations, beginning with a hypothesis which is proved or disproved through study.
- Make a reference or application of this to 'human behaviour' – explanations or models for what can be observed in individuals emotionally and physically in particular situations.

Paragraph 1: Positive aspects (or in introduction)

Give a brief, general overview of the advantages of scientific theory – this should not be the main focus of the essay, but you may wish to allude to why theories that are 'scientific' are often preferred for example:

- Well tested
- Ability to make predictions and apply to other situations
- Indicate possibilities for further research
- Practical applications e.g. using theories to guide treatments for mental illnesses, guiding business strategies, advertising, etc.

Paragraph 2: Limitations:

You should select a few limitations and expand on each in detail, with examples. Alternatively, select more limitations in less detail, but be careful not to just list the limitations and give a brief description at least of each. Points to consider

- GENERALISABILITY is a major problem – can we generalise human behaviour?
- Reductionist/ too simple: It may not be possible to create a unified model of all situations – there are generally far too many factors involved to accurately predict one situation.
- It may not be possible to even identify all factors contributing to a behaviour– conscious and unconscious influences.
- Individual differences in behaviour.
- Should we be focussing on the differences, rather than the similarities, between human behaviour? How far can generalised accounts take us?
- Ethical considerations e.g. free will – people don't like their behaviour to be able to be predicted by a scientific 'formula'.
- It may be dangerous to emphasise particular aspects of behaviour – may focus wrongly on the wrong aspects, causing certain important aspects to be neglected.
- Is it necessary to try and understand behaviour?
- Limitations of our mind/understanding - Is it possible to understand behaviour or are we aiming towards something unachievable? Will we ever be able to understand ourselves?
- Limitations of scientific research – e.g. practical, not all hypotheseses can be tested experimentally, external validity of the findings.

Conclusion:

- This essay question is so broad that really, you can consider any limitations you wish, so long as sufficient argument/ examples/ explanations are given to back up your point.
- The conclusion should briefly summarise the benefits and limitations of scientific theories.
- You may wish to reach a conclusion as the extent to which scientific models of behaviour should be used.

4. Assuming time travel was possible, could we learn more from the past or the future?

Introduction:

- Briefly consider what we might <u>want</u> to learn from looking at the past or future e.g. technology, political decisions, evolution etc.
- Introduce a summary of arguments for both sides of the debate – the benefits of learning from the past, versus the benefits of learning from the future.

Benefits of looking at the past:

- Preventing past mistakes being repeated in the present.
- Understanding previously worked strategies can be used to adapt and produce new strategies to current situations.
- Learning from the knowledge of others requires other people to have experienced things before us.
- History will never repeat itself, but the future we will eventually reach.
- Looking at the past gives us the ability to change the future, but looking at the future does not provide an opportunity to change the past.
- Without an understanding of the foundations/principles of where things came from, we will never be able to fully understand the way things are today, or improve on them.
- A lot can be learned from the way that people lived without the technology, resources, accessibility that we have today.
- Looking at the past makes us appreciate what we have now a lot more.

Benefits of looking at the future:

- If we know what will happen in the future and its demands, we can take steps to be prepared for it e.g. climate change, or change it.
- Again, learning from future mistakes may allows us to prevent them being made so that we can adapt more effectively, knowing what will work and what won't.
- In the future, technology and understanding will advance – therefore if we can look to the future, we can find this out now.
- Focussing on the past leads to regret and guilt, while focussing on the future relieves anxiety and provides an opportunity for things to be changed before they happen – looking at the past can't stop what happened.
- There are endless possibilities in the past, but the future is restricted.
- You may wish to consider ethical implications – briefly consider *if* we should look into the future - how far would this take us?

Conclusions:

- Summarise both sides of the argument – the benefits of looking at the past and of looking at the future.
- You may wish to consider ethical/philosophical debates about looking into the future.
- Arguably, the future becomes the past if you travel to it and then back – so are we ever learning from the future?
- You should come to a final conclusion/summary about what is the most adaptive strategy, if it were possible, since the question asks for a decision to be made.

END OF PAPER

Final Advice

Arrive well rested, well fed and well hydrated

The TSA is an intensive test, so make sure you're ready for it. Ensure you get a good night's sleep before the exam (there is little point cramming) and don't miss breakfast. If you're taking water into the exam then make sure you've been to the toilet before so you don't have to leave during the exam. Make sure you're well rested and fed in order to be at your best!

Move on

If you're struggling, move on. Every question has equal weighting and there is no negative marking. In the time it takes to answer on hard question, you could gain three times the marks by answering the easier ones. Be smart to score points- especially in section 2 where some questions are far easier than others.

Make Notes on your Essay

You may be asked questions on your essay at the interview which might take place upto six weeks after the test. This means that you **MUST** make short notes on the essay title and your main arguments after you finish the exam. This is especially important if you're applying to Economics or PPE where the essay is discussed more frequently.

Afterword

Remember that the route to a high score is your approach and practice. Don't fall into the trap that *"you can't prepare for the TSA"*– this could not be further from the truth. With knowledge of the test, some useful time-saving techniques and plenty of practice you can dramatically boost your score.

Work hard, never give up and do yourself justice.

Good luck!

Acknowledgements

I would like to express my sincerest thanks to the many people who helped make this book possible, especially the 15 Oxbridge Tutors who shared their expertise in compiling the huge number of questions and answers.

Rohan

About Us

Infinity Books is the publishing division of *Infinity Education Ltd*. We currently publish over 85 titles across a range of subject areas – covering specialised admissions tests, examination techniques, personal statement guides, plus everything else you need to improve your chances of getting on to competitive courses such as medicine and law, as well as into universities such as Oxford and Cambridge.

Outside of publishing we also operate a highly successful tuition division, called UniAdmissions. This company was founded in 2013 by Dr Rohan Agarwal and Dr David Salt, both Cambridge Medical graduates with several years of tutoring experience. Since then, every year, hundreds of applicants and schools work with us on our programmes. Through the programmes we offer, we deliver expert tuition, exclusive course places, online courses, best-selling textbooks and much more.

With a team of over 1,000 Oxbridge tutors and a proven track record, UniAdmissions have quickly become the UK's number one admissions company.

Visit and engage with us at:
Website (Infinity Books): www.infinitybooks.co.uk
Website (UniAdmissions): www.uniadmissions.co.uk
Facebook: www.facebook.com/uniadmissionsuk
Twitter: @infinitybooks7

Printed in Great Britain
by Amazon